T0379380

THIS BOOK IS PUBLISHED UNDER THE AUSPICES OF THE PRINCE ALBERT I COMMEMORATION COMMITTEE
APPOINTED BY SOVEREIGN ORDINANCES DATED DECEMBER 14 2018
AND DECEMBER 3 2020 ISSUED BY H.S.H. PRINCE ALBERT II

The WORLDS of a prince

Prince Albert I of Monaco and his times

WRITTEN BY STÉPHANE LAMOTTE

PREFACE BY H.S.H. PRINCE ALBERT II OF MONACO

FOREWORD BY JEAN MALAURIE

ABRAMS / NEW YORK

Preface

September 2021

It was my wish that this hundredth anniversary of the death of my great-great grandfather, Prince Albert I, should be commemorated in 2022 by a series of events, publications and studies worthy of the remarkable achievements of a man whose experiences have resonated with me since childhood.

This book, combining a geohistorical retrospective with a commemorative tribute, was created to remind us of the sheer scale of the prince's undertakings. It draws heavily on photographs and often-unpublished archival images that represent the best-known aspects of the prince's oeuvre but, like the tip of an iceberg, they also suggest what lies beneath. Most importantly, they shed light on the prince's less obvious achievements, such as his humanitarian commitment in defense of Captain Dreyfus, and his efforts in pursuit of world peace when serving as a mediator between France and Germany.

The originality of this book lies in the approach chosen by the author, Mr. Lamotte: a deliberately geographical approach that takes us on a journey from North America to Eastern Europe and Cape Verde to the Artic. At the helm is a "scholar prince" who over the course of a nearly 74-year lifetime devoted himself to the pursuit of Right, Justice and Truth.

Through his numerous writings and his desire to share and persuade, the prince sought to leave a record of his passage. In recognition of that commitment, this book features eloquent tributes to his legacy from the world of letters–a moving testament that touches me deeply.

May this book reach the widest audience possible so as to broaden knowledge of an ever-expanding field of study, and deliver a message that remains more than ever worth pondering today.

Albert de Monaco

Foreword

BY JEAN MALAURIE

My first encounter with the man who had just been crowned Sovereign Prince of Monaco, Albert II, will always remain one of my most precious and moving memories.

It took place in my apartment in the 1st arrondissement of Paris, opposite the residence of the great French writer Louis Aragon. The young prince suddenly came up to me, plainly curious about his Monegasque people and their long history of sovereign statehood. Anyone could see that he took his new responsibilities to heart, only now becoming aware of just how remarkable were those values upheld by his forebear and namesake throughout his own years as sovereign.

And that, of course, was why he was approaching me now. Because of my role as an active environmentalist at international level, committed to protecting the planet and particularly Thule, the northernmost known spot on Earth. Albert II saw me as a sort of ferryman, someone who could steer him in the right direction. Help him find out more about this illustrious great-great grandfather of his, whose extraordinary maritime and scientific achievements had given the Rock of Monaco international credentials.

That first meeting marked the commencement of Prince Albert's enduring support for France's scientific endeavors and my own research objectives in particular. I am especially grateful to him for having allowed the successive publication of four fundamental research papers, *Arctica I, II, III* and *IV* (CNRS Editions), and for his unflagging sponsorship of the works conducted by the Center for Arctic Studies in Greenland and St. Petersburg.

But I digress. As the author of this foreword, it is my privilege to write about His Serene Highness Albert I, the man to whom this book is dedicated. His outstanding academic legacy is repeatedly attested here and by other sources, all of them recognizing his role as the first Prince of Monaco to have placed Monaco's legendary "rocher" firmly within the domain of ocean science. Thanks to his explorations and those of French naval officer, Jacques Cousteau, Monaco's Oceanographic Museum enjoys uncontested authority in the field of marine research. In the case of Prince Albert, that authority also encompassed Spitsbergen, where his many discoveries shed light on unknown regions and served to reveal his remarkable literary talent.

"Here then is Spitsbergen itself, tucked away in its polar refuge, trampled underfoot by ignorant folk who in the name of tourism vulgarize their fine surroundings through sheer stupidity. To see the wonders of this landscape exposed to the gaze of gawping nonentities fills me with sorrow. It is as like seeing a lovely woman being defiled by fools." (*La Carrière d'un navigateur,* 1951 edition, page 298).

"I love the North because death passes by with the dignity of silence, quietly interring in the crystal expanses of the ice fields those who worldly lies have destroyed." (*La Carrière d'un navigateur,* 1951 edition, page 268).

Yes, Prince Albert I Sovereign Prince of Monaco was a poet and a writer, a man whose vision encompassed exploration from the Mediterranean Sea to the North Pole. He must also be remembered as a prince of the Grimaldi dynasty who played a decisive role by asserting Monaco's independence vis-à-vis Paris, London and New York. Yes, the Principality of Monaco had an unassailable claim to sovereignty and he proved it by rallying to the support of Captain Alfred Dreyfus at the height of the crisis that rocked France. The prince suggested that Dreyfus stay at the Grimaldi's Marchais property on his return from the penal colony of Cayenne, a proposal that saw him branded a "Jew" and "Germany's lackey" by France's right-wingers. It was thanks to Albert I that Monaco became a truly independent, sovereign state.

The Principality of Monaco now seeks to go further still, by establishing itself as a global hub of forward thinking, most notably about the environment and the Arctic environment in particular. It is without hesitation that I donate the entire body of my Arctic research, in Greenland, Canada and Siberia alike, to make that project a reality. The Russian authorities in St Petersburg together with the Polar Academy join with me in placing the Jean Malaurie Institute of Arctic Studies, headed by the very learned Professor Jan Borm and closely associated with the Prince Albert II of Monaco Foundation, under the authority of the University of Versailles Saint-Quentin-en-Yvelines, an associate member of the Paris-Sarclay University. It is incumbent upon us all–me, my successors and my fellow French citizens–to rise to this occasion and help build Monaco's reputation as a world forum for polar research, in the manner of Geneva as a world peace forum and Stockholm as a world Nobel Prize forum. It is my dearest wish that the spirit and forces of the Polar World may help us live up to the international vision of His Most Serene Highness Prince Albert I of Monaco in pursuit of an undertaking embodied and expanded by the new UNESCO programs.

As the representative of the Centre National de la Recherche Scientifique (National Center for Scientific Research, CNRS) and the École des Hautes Études en Sciences Sociales (School of Advanced Studies in the Social Sciences, EHESS), I earnestly salute Monaco's wishes to enhance its scientific and moral credibility through the acquisition of my Arctic collections by its much-celebrated museum. It is a matter of urgency to draw up a program, conducted under the auspices of the Jean Malaurie Monaco-USVQ Arctic Research Institute, to help protect the Earth and the largely forgotten indigenous Arctic Peoples. The first priority is to defend their climate, ecology and way of life, sparing no effort to protect these Peoples of the Great North whose future is somehow entwined in the Apollonian destiny of one of the most valuable natural places on Earth.

Scientific rigor is a vital adjunct to the precautionary measures taken by Western countries to prevent any acceleration of a worrying and hitherto unchecked history of climate change.

Contents

1848
1922

"We dance like a cork on an ocean of mad waves that overtake us from moment to moment."

Julien Gracq, *Le Rivage des Syrtes*, 1951.

ews of the death of Prince Albert of Monaco will be received with profound sadness not only in the scientific circles that held him in such high esteem but also by all of those people who were dear to his heart and counted themselves as his friends." (*L'Intransigeant*, June 29 1922). So wrote a French daily newspaper following the demise of Prince Albert I of Monaco on June 26 1922–warm words for a newspaper that was not always kind to the prince in his lifetime. Tributes flowed from near and far. On July 30 that year the neighboring commune of La Turbie (southeastern France) christened a new road after the prince. In October, Norwegian military officer and polar scientist Gunmar Isachsen wrote in *Naturen* that "the prince did more than any other foreigner to further the advancement of Norwegian scientific research." Heartfelt testaments like these give an indication of the sheer scale of the prince's works, a man who people saluted as a sailor, a scholar, a statesman, a humanist and a patron of the arts. Understandably therefore, the principality has regularly sought to revive the memory of such an emblematic figure in its history.

In Monaco, the story in remembrance of Prince Albert I has unfolded on several occasions. It began in the aftermath of World War II, with the commemoration of the hundredth anniversary of the year of the prince's birth, on Saturday November 13 1848. The celebrations revolved around three symbolic places: the Oceanographic Museum of Monaco founded by the deceased as a temple of the sea; the cathedral and the special service in honor of the prince; and *La Place du Palais* where a tribute ceremony took place in front of *La Science découvrant les richesses de l'Océan*: a monument to science and the sea, unveiled in 1914.

The official speeches predictably presented the prince in the guise of a scholar and a navigator–a new icon for a proud and grateful people no doubt anxious to assuage the stigma of the recent hostilities.

In the reign of his son Prince Louis II, his father was too often remembered in hagiographic tones that blurred the line between reality and fantasy. Within a mere 26 years of his death, all that remained of Prince Albert I were a collection of ghostly personas. "Mathematician, physicist, navigator, diplomat, sportsman, mechanic, administrator and statesman"–the prince became all of these things and all of them were true. But so too were other things that got lost amid the heartfelt accolades–the worshipping of the hero at the expense of the man himself.

View of the oceanographic museum in 1913, three years after its inauguration.

The years that followed brought other little reminders of the prince: the statue by sculptor François Cogné in the Saint-Martin gardens, facing the Mediterranean (1951); a new edition of his book *La Carrière d'un navigateur* (The Career of a Navigator, reissued in 1951 and 1966, first published in 1902); and the rechristening of the Monaco Lycée, henceforth called the Lycée Albert Ier in homage to its founder (1960). The fiftieth anniversary of his death in 1972 then passed relatively uncelebrated and it would take until 1998 for the past to resurface, this time on the occasion of the one hundredth and fiftieth anniversary of his birth, with the spotlight once again on 1848–the same year that neighboring France celebrated the triumph of the Second Republic. This was in the reign of Rainier III, the "builder prince" who put the principality firmly on the map. With the benefit of hindsight, Rainier sought to depart from hagiographical representations of the prince's life and work and subject them instead to a fearless critical appraisal: Jacqueline Carpine-Lancre's *Albert Ier, prince de Monaco. Des œuvres de science, de lumière et de paix*. Published in 1998 to coincide with the remembrance celebrations, the book opened the way for a wealth of source material that gave emphasis to the prince's writings (correspondence, log books, scientific papers) while also revealing the complexity of his life–his pursuit of balance between the ideal of an organic society and the fear of degeneration.

Monument in memory
of H.S.H. Prince Albert I,
by François Cogné,
St Martin Gardens, Monaco.

And so we come to 2022. Another centenary. The year when several centenaries come together, weave links, create meaning: the centenary of the death of Belle Epoque writer Marcel Proust; the centenary of the death of British explorer Ernest Shackleton; and the centenary of the birth of pioneering French scientist, Louis Pasteur. Albert I takes the podium alongside them, a place that he must be shown to deserve.

This is the first time Monaco will be celebrating the anniversary of his passing, focusing on death as the consummation and not the end of his life. A life lived to the full that bequeathed an enduring legacy. The work of a lifetime condensed into a single emblematic moment at the turn of the 19th century–the moment that marked the completion of Albert I's 33-year-long reign. The year 1922 must be seen as a date alive, a landmark date, not a death date engraved on a headstone. The celebrations are long awaited, most especially because their organization is entrusted to a special commemorative committee appointed in 2019 by His Serene Highness Prince Albert II of Monaco. Their task is to assemble a program of events that serve as a showcase for the principality–that reveal modern-day Monaco through the prism of Prince Albert I's humanism and internationalism. The aim is to represent something of the past in the present.

S.A.S. LE PRINCE
ALBERT 1er
1848-1922

LEFT-HAND PAGE
Sculptor François Cogné poses alongside his work unveiled on April 11 1951.

RIGHT
Monaco as it was in 1913.

From top to bottom:
Monte Carlo: La Condamine as the port restructuring nears completion;
Le Rocher: Prince's Palace, Oceanographic Museum and Hôtel du Gouvernement;
New Districts: Les Salines, hospital and cemetery. Fontvieille, new industrial zone and municipal solid waste incineration plant.

Memory as political strategy certainly, but a methodical approach nonetheless that makes it possible to appreciate the "otherness" of the past and construct a relevant historical discourse, helped by the remoteness of the subject under study and the conspicuous absence of Prince Albert I himself. The intention is a broader sweep, placing emphasis on the international and political dimension of the sovereign's works, a man for whom science and civilization went hand in hand: "[...] we will continue to give the best of ourselves to the development of scientific knowledge, in order to provide civilization with a firm foundation and raise human nature above the troubles that issue from mankind's obscure origins," (Rome, conference on progress in oceanography, April 27 1910). Whether at sea, in his palace, in Rome or in Paris, Prince Albert I governed in every sense of the word. He was a Man of Action, driven by an ideal of peace that he upheld in support of two cherished causes: the vindication of Captain Dreyfus, and pacifism as a vehicle for avoiding World War I. He believed that science and politics should serve humankind, furthering human welfare at all times and in all places. It is in recognition of these universal values upheld by the prince in his role as an arbitrator that Monaco's neighbors and friends are now showing renewed interest in his legacy–the mark he left in places of collective memory, most notably in Italy, Spain, France, Portugal, Norway and Germany.

The prince dressed for the weather on board the second *Princesse-Alice* in 1907.

As this book sets out to show, the course pursued by Prince Albert must be understood in the context of geography and not in terms of those "harder" sciences traditionally favored in an account such as this. This is the story of the geography of a journey, a vagabond journey since it took place at a turning point in history: the point between the last great journey of exploration–the two poles were reached between 1909 and 1911–and the tentative beginnings of mass tourism. The point when the world began to shrink as people came to know it better. Prince Albert I was born in a century that was eager to find out about the world. Prince, voyageur, inquisitive explorer, intellectual, representative of the nation, passionate hunter, frequenter of spa baths–Albert I was all of these things. Only the pilgrim was lacking. Ever on the move as if to escape, the prince drew attention to himself by his seemingly ubiquitous manner of being in the world. His geographical leanings were nurtured by a childhood spent devouring the great travel stories that flourished in the late 19th century. The *Hirondelle* was to Albert I what the *Beagle* was to Darwin before him. But travel brought obligations too, as embodied in Albert's inescapable duty to undertake the Grand Tour.
In between visiting Marchais, Orléans, Paris and Monaco, the young prince longed to broaden his horizons, partly no doubt to shake off the shackles of status and escape a preordained future that left him no room to maneuver. In the meantime, he journeyed into himself, confronted his doubts and melancholic thoughts and succeeded in stifling his dream, displaying a tenacity that

makes him more akin to Pierre Aronnax than Captain Nemo. Travel stories made way for more learned reading matter that provided him with a solid fund of knowledge even if it wasn't acquired on university benches. His was of course a personal geography, complete with map coordinates of his chosen landscapes: those "archipelagos of passion," the Azores in Portugal and Spitsbergen in Norway. These two countries as a whole stood at the center of his geography of emotion and knowledge. There is also the fact that vagabonding strips away unnecessary baggage, forges a personality and sharpens perception of the world. As Albert discovered firsthand, it also debunks the myth that travel transforms a sailor. Travel, he realized, was a lesson in simplicity, a chance to cultivate heroism in everyday life by refusing entitlement. A meal shared with his sailors or the inhabitants of the islands of the Azores was far removed from the fawning and stultifying formality of ceremonial protocol. Going down into the abyss was also his way of challenging certainties in search of self-knowledge. Henceforth it was the encounter with otherness and the results of the expedition that justified the journey. Those results were published from 1889 onward in the *Résultats des campagnes scientifiques accomplies sur son yacht par Albert Ier, prince souverain de Monaco* (Results of the scientific campaigns carried out on his yacht by Albert I, Sovereign Prince of Monaco). His approach must also be considered in the broader context of the drive to promote geography as a patriotic undertaking–in France, in the years following the defeat of 1870, and in Germany in the years following unification. But when Albert I joined the Société de Géographie de Paris in 1885 he had a different type of patriotism in mind. Through scientific campaigns essentially supported by the foremost French, British and German scientists of his times, he showed solidarity with Europe and raised awareness of his little country. The Principality of Monaco came into existence in the course of talks held at the palace and the university. The winds that filled the sails of the *Hirondelle* served to chase away the bad image of the casino, that temple of the dark arts vilified by France's far-right press. Through his many journeys and his openness to the world, his calls for solidarity as an antidote to identity politics, he came to invent a story about his nation that still today spreads the reputation of the principality far beyond its minuscule territory. Such was the singular achievement of this prince, poised between the desire to leave himself behind and the imperious necessity to uphold his position. He would also come to love and protect the places he studied, his twofold concern for the origins of life and its preservation marking the culmination of a rich program of geographic endeavor committed to sustainability. His works echoed those of Alexander von Humboldt and Élisée Reclus: "The urgency of the moment is upon us: faced with the threat of industrial activity we must make every effort to prevent the destruction of the remnants of organic life buried beneath our feet by the passing of time [...] That is why the annihilation of animal and plant species bodes ill for the future in the same way as the loss of the records stored in the paleontological archives of our planet." (AMOM, manuscripts of Prince Albert I, Paris, January 1914).

"A flurry of ghosts suddenly emerged before my eyes." Chapter I "The soul of a sailor." Engraved illustration by Ernest Florian from a drawing by Louis Tinayre, *La Carrière d'un navigateur*, 1913-14 edition.

Following this apercu of the milestones in the prince's career, at a time when the climate crisis and the coronavirus pandemic raise issues for freedom of movement, the moment has come to look at those journeys he undertook–some ordinary, some extraordinary, but all of them undertaken in a bygone time redolent of otherness. The first stage quite naturally takes us to Monaco and France, featuring documents and photographs, some of them never before published, that together offer a new way of representing the world that Prince Albert I helped to create.
As we move through the prince's peregrinations, the episodes rely heavily on illustration, the aim being to put together a book-cum-logbook–a record for posterity that also serves to catalog the exhibitions held to mark the prince's centenary.
This book is intended as an accompaniment to *Journal de ma vie*: a flagship publication comprising a collection of the prince's handwritten diaries, to be published simultaneously by Editions Perrin. Essential to a proper understanding of the prince's life and works, *Journal de ma vie* is a clear demonstration that the politics of memory have pedagogical value. They serve to encourage and enrich editorial possibilities, bringing to light works that might otherwise remain confined to a handful of specialists, and also breathing new life into the historical approach.

"The urgency of the moment is upon us: faced with the threat of industrial activity we must make every effort to prevent the destruction of the remnants of organic life buried beneath our feet by the passing of time [...] That is why the annihilation of animal and plant species bodes ill for the future in the same way as the loss of the records stored in the paleontological archives of our planet."

Prince Albert I, 1914

PREVIOUS PAGES
The prince's yacht, the second *Princesse-Alice*, in the port of Monaco, 1909.

LEFT-HAND PAGE
The prince in the final years of his life, lost in thought.

MONACO and France

Homelands of the heart and of power

"And there is a special pleasure for some minds
in the reflection that we share the impulse
with all outdoor creatures in our neighbourhood,
that we have escaped out of the Bastille of civilisation,
and are become, for the time being, a mere kindly
animal and a sheep of Nature's flock."

Robert Louis Stevenson
***Travels with a Donkey in the Cévennes*, 1879.**

It may seem surprising to combine France and Monaco in the same chapter when one considers that no-one did more than Albert I to win recognition of the independence of his little state. But then again, the prince regarded France as his second homeland. Perhaps it would be more appropriate to talk of a *petite patrie*–a little homeland where he put down roots as a way of extending a domain that may not have fulfilled his expectations of grandeur. Whatever the case, it was in France that he set up the network to serve his political purposes, drawing support from academic circles that ensured his works would resonate with the country's many learned societies. It was also there that he surrounded himself with the men he needed to devise a Rule of Law and Absolute Sovereignty. Hence the many places in France where he left his mark, especially Paris, Marchais and the ocean-facing ports, and even more so the Plateau du Carladès and the Pyrenees in the interior of the country. Together they create a picture of his personal geography, serving as landmarks within his emotional maps.

PREVIOUS PAGES
Terrace of the Monte Carlo Casino, with Cap Martin in the background.

ABOVE
Photo of Prince Albert as a teenager, by A.A.E. Disdéri. Photo colorization by Parfu (undated).

RIGHT-HAND PAGE
Prince Albert I on board his yacht in the port of Monaco. Photo by Charles Chusseau-Flaviens, 1905.

ABOVE
Engagement celebrations of Prince Albert and Mary Victoria Douglas-Hamilton, 1869.
Photo by Lejeune.

RIGHT-HAND PAGE
Pierre-Henri-Théodore Tetar van Elven, *Fête de Nuit aux Tuileries, le 10 juin 1867, à l'occasion de la visite des souverains étrangers à l'Exposition Universelle* (party at the Tuileries, on June 10 1867, to mark the presence of foreign heads of state at the World Fair.)

Like so many of his predecessors, it was in Paris that Albert Honoré Charles Grimaldi came into the world, in the apartment rented by his parents Prince Charles III and Princess Antoinette-Ghislaine at 90, Rue de l'Université. It was also in Paris that he spent his youth and gained a solid grounding in how to run a country, first under the watchful eye of his tutor, l'Abbé Theuret, then at an institution in Auteuil followed by the Collège Stanislas and later the Petit Séminaire d'Orléans. It was in Paris too that he laid the foundation for his future network and began to dream of adventure and epic wanderings inspired by travel literature. As the boy grew to manhood, Paris came to signify imperial celebrations, Haussmannian transformations and a cultural dynamism mingling avant-gardism with controversy. It was a place of lessons learned in the salons and soirees hosted by Marie-Lætitia Rattazzi, and information gleaned from circles of politicians, writers and businessmen on the rules of protocol and the codes and arcana of power. It was the place where he met his first wife, Mary Victoria Douglas-Hamilton, who was introduced to him at Les Tuileries by Napoleon III himself. In his pre-pacifist days as a sign of loyalty to the Emperor of the French, the young hereditary prince served in the French Navy and took part in the campaign against Prussia.

P. Van Elven

After the fall of the Second Empire, he adapted perfectly to the social imperatives of the new regime, as evidenced by his attendance at the state funeral of Victor Hugo in 1885. Defying etiquette and expectations, the young prince of Monaco elected to share in a moment of the purest Republicanism in order to "glorify one of the finest manifestations of intelligence." His presence at the funeral is eloquent testimony to his pre-occupation with presence and standing center stage. Another civic event linking him to France, this time of a more ritualistic nature, was the Paris World Fair of 1878 and especially its 1889 and 1900 incarnations.

ABOVE
The Eiffel Tower, emblem of "Belle Epoque" Paris.

RIGHT-HAND PAGE
Victor Hugo's funeral, showing the catafalque under the Arc de Triomphe, May 31 1885; and the crowd gathering behind the cortège, June 1 1885.

THE PARIS WORLD FAIRS OF 1889 AND 1900

It was not enough that he should merely represent the principality; the prince seized the opportunity to present the results of his scientific campaigns in the two consecutive Paris World Fairs of 1889 and 1900, where his collections won several medals. His presence there was in line with his interest in museums, most notably the National Museum of Natural History where his discovery in 1884 of the results of the campaigns of Alphonse Milne-Edwards aboard the *Travailleur* and the *Talisman* cemented his vocation as an oceanographer. He was also an active member of several learned societies, among them the French Society of Geography (1885) and the French Academy of Sciences, which he joined as a Corresponding Member in 1891 before being appointed a Foreign Associate Member in 1909. It was this thirst for truth as an antidote to doubt, his fears of dissolution and his horror of fin de siècle decadence that drew him to the faith in progress espoused by the neighboring French Republic. Following his attempts to popularize the results of his scientific campaigns, in 1906 he founded two major Paris institutions intended to make science accessible to everyone:

the Institut Océanographique (established in 1906); and the Institut de Paléontologie Humaine (established in 1910). The latter institution was a further demonstration of his commitment to action, reflecting his beliefs that the origins of life should be studied across time and space. It also testified to his determination as a man of science to leave his stamp on France, and in the process enhance Monaco's reputation by giving scientific research an international dimension.

His presence in Paris was made even more emphatic by what for a sovereign prince were unusually firm political convictions. He was, for instance, on friendly terms with French presidents Félix Faure, Armand Fallières and Paul Deschanel, who he received at his palace in Monaco and visited at the Elysée. In 1894 he also met with President Sadi Carnot, followed by Émile Loubet in 1901. Then there were his meetings with President Félix Faure to plead the case of Captain Dreyfus, in Nice and in Monaco in 1896 then again at the Elysée in 1899 shortly before the president's sudden death. His Dreyfusard sympathies did not go down well with the French nationalist press. With their usual gift for confusing the issue, right-wing newspapers decried the interference of a foreign prince with links to the infamous casino, meanwhile raising fears among the Dreyfusards that the image of the House of Grimaldi might damage their cause. But there were also those who saluted the prince's courageous stand, among them close associates such as Flore Singer and Joseph Reinach; also Émile Zola, the Countess Greffulhe and of course the Dreyfus family themselves, Lucy and Alfred alike. All of them were grateful to the prince for his swift reaction, sincerity and informed commitment. While at sea on course for Svalbard, the prince took advantage of a stopover to read the French press and catch up on the ins and outs of the affair. In Monaco, he welcomed fellow Dreyfusards, Commander Ferdinand Forzinetti, Commissaire Tomps and the Abbé Pichot, when they sought refuge in the parish of Sainte-Dévote in 1900. His unswerving commitment to pacifism on the eve of World War I is meanwhile evidenced by his attempts to bring France and Germany closer together by acting as host for the Universal Peace Congress in 1902. It was also the prince's intention that the conference should serve to make Monaco a hub of international arbitration,

LEFT-HAND PAGE
The Monaco Pavilions at the 1889 and 1900 World Fairs in Paris. Two prime locations: in 1889 on the Champ de Mars, close by the highlight of the exhibition; then in 1900 on the Rue des Nations, along the River Seine.

ABOVE
Entrance ticket for the 1889 World Fair.

like Berne, Brussels or Le Havre–a wish fulfilled the following year when the International Institute for Peace took up residence in the chapel of the former Hôtel Dieu in Monaco. Its director and trusted advisor Gaston Moch would play an important role here, as would the prince's epistolary friend, Baroness Bertha von Suttner, a leading figure in the international peace movement and the first woman to win the Nobel Prize for Peace in 1905. The prince's pacifism however should not be confused with passivity. At the outbreak of war in 1914 he put the Paris institutes of oceanography and paleoanthropology at the disposal of the French government, together with the wireless telegraphy aboard his yacht. He offered the French Red Cross use of the Chateau de Marchais and the Monaco Hospital; and in 1915 he is believed to have taken advantage of his activities at sea to help the French Navy retrieve munitions and gunpowder sunk off the coast of Toulon. In September 1914 meanwhile, following the destruction of Reims Cathedral, he had sent a telegram to President Raymond Poincaré expressing solidarity and outrage. Combined with the damage caused by the German bombardment of Paris to his private residence at 10 Avenue du Trocadéro (rechristened the Avenue Président-Wilson in 1918), this led to a severing of all ties with Wilhelm II.

BELOW
Official opening of the Institut International de la Paix in Monaco, February 25 1903. With the prince are Bertha von Suttner, Gaston Moch, Abbé Louis Pichot, Gustave Saige, Charles de Monicault, Henri de Maleville and Edmond Izard.

RIGHT-HAND PAGE
Extract of a letter from Prince Albert I to Alfred Dreyfus, written on board the second *Princesse-Alice* as the prince sailed to Spitsbergen. Dated July 10 1899, it was composed shortly before the opening of the retrial in Rennes. Pictured right is a photo portrait of Prince Albert I.

En mer 10 juillet 1899

(une variante a été envoyée de Tromsö le 14 et datée le 15)

Au capitaine Dreyfus,

epuis longtemps j'affirme votre innocence
arce que je la vois comme je vois la clarté
u jour. Vous même en vrai soldat vous
affirmiez mieux que par des serments
orsque vous traversiez debout le champ
u supplice.
ussi les braves gens sont avec vous, et
si des aveugles persistent à s'insurger
contre le sentiment qui anime l'élite de la
nation, ils s'exposeront, j'en ai la certitude,
à d'humiliants démentis.
Mais les juges suprêmes que vous avez
demandés sont là, et ils savent que si
l'épée anoblit l'homme qui la porte,
c'est seulement lorsqu'elle symbolise
dans sa main la droiture et la force.
Ils savent que leur nom va contresigner
définitivement le verdict d'où sortira
pour la France une nouvelle grandeur
faite de ses angoisses et de sa loyauté.

paraissent infimes, que je parle ainsi de
justice et de réparation.
Et puis la force que prête aux consciences
l'habitude des grands problèmes du monde
me donne une vision de la France, de
celle qui enveloppa les peuples de son
génie civilisateur, et qui acclame aujour-
d'hui des soldats et des juges dans un
élan dont les siècles garderont la mémoire.

Albert Prince de Monaco

Dans l'attente de la parole libératrice des
juges, les âmes qui ne sont ni de pierre
ni de boue admirent l'inévitable destin,
en vous voyant ramené par la tempête
qui chasse les mensonges et les crimes.
Elles voient prochains la fin d'un triste
rêve, le dénouement qui montrera
comment une armée consciente de son
honneur rouvre ses rangs au frère
d'armes grandi par la victoire de la
justice.
Loin du lieu où les plus clairvoyants de
[illegible] accomplissent envers vous
[illegible] très haute et

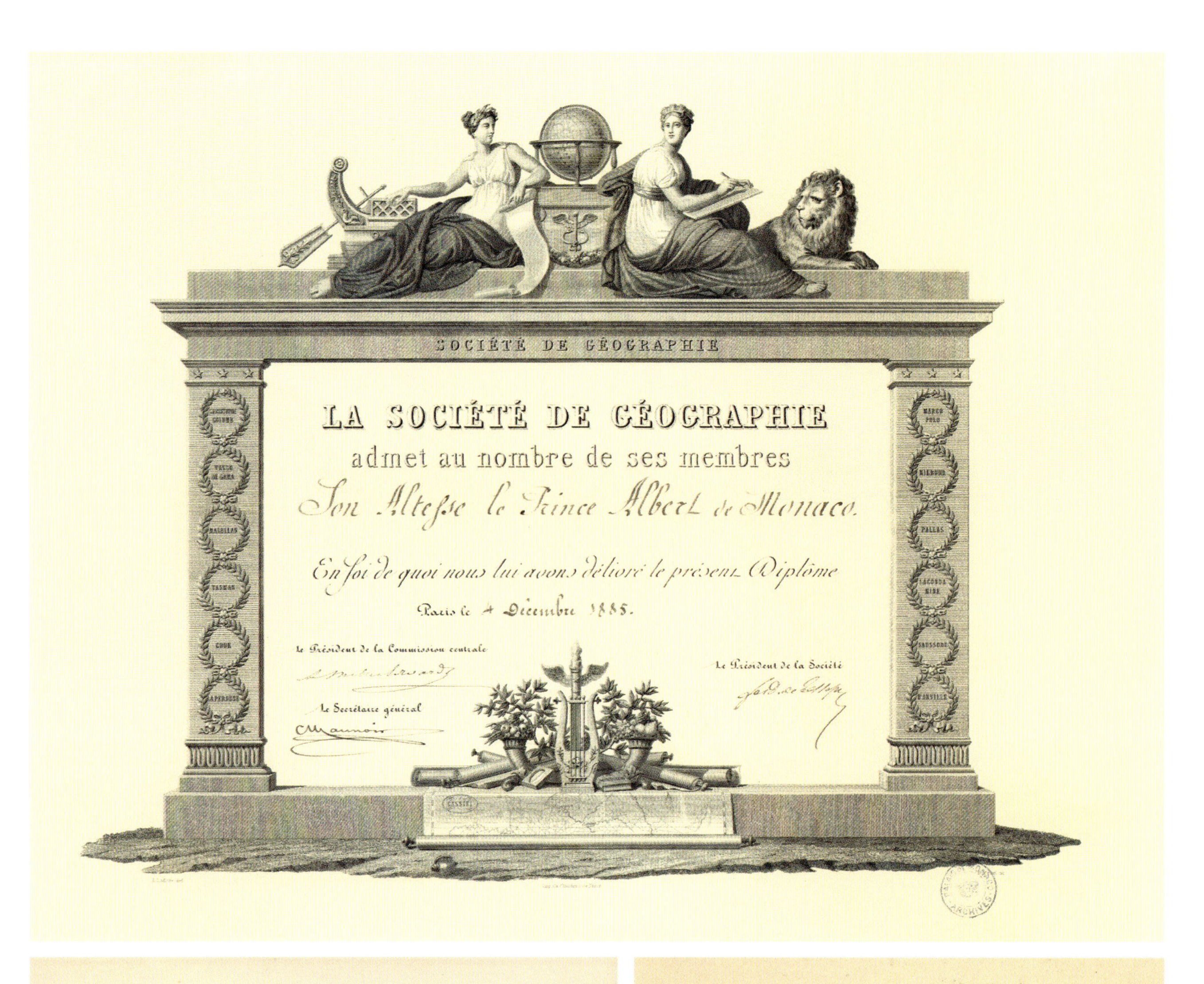
SOCIÉTÉ DE GÉOGRAPHIE

LA SOCIÉTÉ DE GÉOGRAPHIE
admet au nombre de ses membres
Son Altesse le Prince Albert de Monaco.

En foi de quoi nous lui avons délivré le présent Diplôme

Paris le 4 Décembre 1885.

Le Président de la Commission centrale

Le Secrétaire général

Le Président de la Société

INSTITUT DE FRANCE

ACADÉMIE DES SCIENCES

Paris le 29 Mars 1909.

Les Secrétaires perpétuels de l'Académie
à Son Altesse Sérénissime Albert Ier, Prince de Monaco,
Associé étranger de l'Institut de France.

Monseigneur et très honoré Confrère,

Nous avons l'honneur de vous informer que, dans la Séance de ce jour, l'Académie des Sciences vous a nommé à la place vacante parmi ses Associés Etrangers par suite du décès de Lord Kelvin.

Aussitôt que l'Académie aura reçu

EXPOSITION UNIVERSELLE DE 1900

PRINCIPAUTÉ DE MONACO

LES

CAMPAGNES SCIENTIFIQUES

DE

S. A. S. LE PRINCE ALBERT Ier DE MONACO

PAR

LE Dr JULES RICHARD

Chef du Laboratoire de la *Princesse-Alice*

Conservateur des Collections scientifiques de S. A. S. le Prince de Monaco

IMPRIMERIE DE MONACO

1900

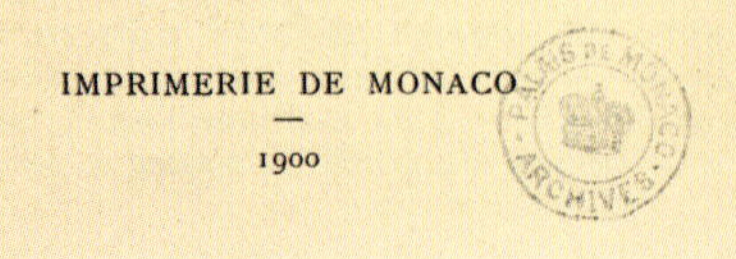

THE HAVEN OF THE CHATEAU DE MARCHAIS

The prince's resentment against Germany was no doubt sparked by the damage caused to his Picardy estate, the Chateau de Marchais, while under occupation by the Germans. Without Marchais, Paris meant nothing to the prince. Marchais was in his blood. It was the place where he had learned the meaning of life and felt the first stirrings of passion; the place purchased in 1854 by his mother, Antoinette-Ghislaine de Monaco, born Comtesse de Mérode, partly because of its proximity to her native Belgium. Marchais conjured up memories of idyllic childhood days spent in play or satisfying his appetite for sports. It was the little Garden of Eden that had fed his curiosity about Nature and forged his identity in the process. It was also the place of long walks with his mother, listening to her concerns about the plight of the poor and developing a social awareness that would later show through in the interest he showed towards the Mutualité Française (the national federation of mutual aid associations, founded in 1902). When he grew to adulthood, Marchais was the place where he recharged his batteries and first dreamed of adventures at sea. It would always be a fixture in his life: a retreat, a refuge in moments of doubt and the place where he hosted hunting parties and visits by foreign royalty, scientists and other dignitaries. But driven by his insatiable appetite for discovery, he also turned the property into a venue for research, enriching and reshaping the woodlands abutting the commune of Sissone and even placing the chateau itself at the service of science. In pursuit of his particular interest in wireless telegraphy, he invited engineers onto the property to demonstrate the transmission of electrical currents through the earth without wires, and to carry out experimental

LEFT-HAND PAGE
Scientific credentials from Paris: certificate of admission to the French Society of Geography, December 4 1885; letter to the prince confirming his election as a Foreign Associate Member of the Paris Academy of Sciences, March 29 1909; publication of the results of the prince's scientific expeditions on the occasion of the 1900 World Fair.

ABOVE
The Château de Marchais in the Aisne department, bought by Antoinette-Ghislaine de Monaco in June 1854.

studies on soil improvement under the supervision of the *Grandes Écoles* of agriculture. Marchais became one great workshop for the production of new ideas–an arena for the meeting between politics and science where guests never quite knew who they might meet. In the autumn of 1898, for instance, the enthusiasm for moving pictures overriding good taste, the talented but highly controversial French surgeon Eugène Doyen was invited to Marchais to present films of his operations.

The other places dear to Prince Albert I were scattered here and there and are too many to mention in these pages. But a few deserve special mention, starting of course with the neighborhood of La Turbie, on the doorstep of Monaco, where the prince helped to restore the Trophy of Augustus. Also Roquebrune, home of his friend the Empress Eugénie, and further afield, the port towns of France's Atlantic coast where his longing for distant shores took form in the hiring of sailors and preparations for departure. Without those towns, there would have been no scientific expedition. Towns like Lorient in Brittany, where he began his naval career at the officer training school; and Cherbourg and Le Havre in Normandy, which acquired particular significance as the jumping-off points to the Portuguese and Norwegian islands. The prince's crews mainly consisted of Breton sailors whose skills and devotion he greatly

BELOW
Prince Albert I on his Humber Beeston motorcycle in the Palace gardens, March 1905. Photo by Charles Chusseau-Flaviens. One snap in this series made the front page of the sporting periodical *La Vie au Grand Air* (the outdoor life) on March 30 1905.

RIGHT-HAND PAGE
"Monaco. Monte Carlo." Poster by Alfons Mucha, PLM, 1897.

MONACO·MONTE-CARLO
CHEMINS DE FER P.L.M.
Billets d'Aller & Retour_Billets Circulaires
Billets d'Aller & Retour collectifs de famille à prix réduits.
TRAJET EN 16 HEURES, PAR TRAINS DE LUXE
Mucha
AFFICHES ARTISTIQUES DE LA SOCIÉTÉ "LA PLUME"
31, Rue Bonaparte, PARIS.
IMP. F. CHAMPENOIS.
66, Boulᵈ Sᵗ Michel, PARIS.

appreciated. His sailing ship the *Hirondelle* meanwhile overwintered in Lorient–a fact endlessly trumpeted by local journalists for whom Prince Albert I was clearly of Breton extraction. Then there was the string of places he visited in his wanderings on two wheels, what he called his own "Tour de France": a tour by stages undertaken in the three years from 1902 to 1905, riding a Clément *autocyclette* (motorized bicycle) or a Humber Beeston motorcycle, with stops between Paris and Marchais then Monaco and Paris. For this lover of the sea, being anchored in the earth could be a salutary experience but also an opportunity for elevation, as for instance when touring the mountains of the Auvergne (the Rock of Carlat) and especially the Pyrenees, for which he had a particular fondness. Between Luchon and the Val d'Aran in southwestern France, he would indulge his passion for hunting but with a growing awareness of the need to preserve and protect the environment, particularly in the last years of his life. In 1917, inspired by his recent visit to the USA, he contemplated the idea of creating a cross-border nature reserve between France and Spain, a place faithful to the spirit of fin de siècle "pyreneism" that regarded the physical experience of the mountains as inseparable from the emotional and meditative experience. For Prince Albert I, journeying appeared without end–a path with no arrival.

Prince Albert I and Princess Alice, married on October 30 1889.

Prince Albert I was every bit as French as his father Charles III and his artist grandfather Florestan before him, perhaps even more so. So what must we make of his relationship with Monaco? Jump to the conclusion that the principality came second for him? That his absences were motivated more by indifference than diplomatic or scientific imperatives? Tempting, but no. Despite the myth that has been circulating for decades, Prince Albert I resided in Monaco, came from Monaco and was devoted to Monaco. Monaco was his home in the spring and winter and also in the intervening months depending on the recommendations of his close aides.

MONACO, ACCESS TO THE SEA

Until the mid 19th century, Menton and Roquebrune were under Monegasque rule. That changed in 1848, the year of the prince's birth, when they voted in favor of being annexed by France, which became official in 1861. Their loss considerably reduced Monaco's territory. But as an adult, the prince would put aside his sadness and make the most of his country's unique status as the only European micro-state that was not landlocked: Monaco is a semi-enclave, surrounded on three sides by France and fronting the Mediterranean Sea. For Albert I that sea would become an ocean, a source of rest and recuperation, the freedom to depart the better to return. For him Monaco was a multiplicity of states across space and time, poised between present reality and distant preoccupations.

The Prince Albert I and French Presidents

Over the course of his 33-year reign (1889-1922) the prince met and entertained friendly relations with eight presidents of the French Republic.

LEFT-HAND PAGE
Top: Conclusion of the 11th Congrès de la Mutualité Française (national federation of mutual aid associations) in Montpellier, March 30 1913. The prince is seen on the theater balcony looking over the Place de la Comédie, in the company of President Raymond Poincaré.

Bottom: Félix Faure visits Monaco on March 5 1896, seen here arriving at the Palace. The previous day, Prince Albert I had visited the French president in Nice.

RIGHT
Top: Prince Albert I and President Poincaré at the 11th Congrès de la Mutualité in Montpellier, March 30 1913.

Middle and bottom: Armand Fallières visiting Monaco in 1909. The president and the prince seen here after visiting the Oceanographic Museum, one year before its opening. Then together again after lunch, in the Palace gardens.

BEAU-RIV

There was the Monaco of his youth, its territorial borders marked with a dashed line, a necessary point of transition before total immersion. The prince first went there in 1849 aged just eight months but would not stay there long, his training as a prince requiring him to complete his education by undertaking the traditional exploration of Europe known as the Grand Tour. His early years were nevertheless marked by a number of symbolic milestones. On May 13 1858 the then-ten-year-old Albert laid the first stone of the casino in Les Spélugues, then known as the casino Élysée-Alberti, which became the Monte Carlo Casino in 1866. In 1874 he moved The *Hirondelle*, the schooner he had just bought in England, to Monaco—a necessary prelude to several Mediterranean cruises that he would undertake departing from the port in La Condamine, some for their cultural attractions, others for diplomatic purposes, others still to indulge his love of hunting. In 1884 The *Hirondelle* became a research ship and Mediterranean cruises made way for expeditions across the Atlantic. Monaco disappeared from view but was ever in his thoughts; and he would return whenever necessary, as in 1887 when he spent several weeks in Monaco assessing the impact of the earthquake that struck the French and Italian Rivieras.

View of the tramway on the Montée du Beau-Rivage (now the Avenue d'Ostende), so named after the former hotel that was much frequented by overseas visitors. The first tramline, Place d'Armes-Saint-Roman, was opened on May 14 1898.

1889, A SOVEREIGN PRINCE IN MONACO

Charles III died in Marchais on September 10 1889. The new sovereign prince arrived in Monaco from Paris in the early morning to the cheers of the waiting crowd. Crown Prince Albert had long been popular with the Monegasques who saw his accession to power as a source of new hope for their little country; they gathered in the *cour d'honneur* of the palace to swear their allegiance. The same warm welcome was extended to the prince's second wife, Alice Heine, an American of French-German descent, born in New Orleans and a cousin of German poet Heinrich Heine. The couple made their entrance into Monaco, where the mayor handed the prince the keys to the city in the time-honored tradition. The fact is that Albert I not only had a presence in Monaco but also made his presence felt. In June that year for instance, he established a Russian consulate in Monaco—a diplomatic initiative that sent a clear signal to the Russians who holidayed in the principality, among them such distinguished figures as Anton Chekhov, Fyodor Dostoevsky, Aleksandr Ivanovich Herzen and Feodor Ivanovich Chaliapin.

The Monte Carlo Casino in 1902.

With Monaco and the Riviera attracting growing numbers of foreign tourists, the prince was well placed to appreciate the importance of internationalism. Building on his father's achievements, he reconstructed his city-state to resemble Paris. Monaco became a city of entertainment, a scaled-down version of Belle Epoque Paris brimming with insouciance and glitterati, a melting pot of nationalities. It became a center for creativity: for musical productions by French composers Jules Massenet, Camille Saint-Saëns and Hector Berlioz; for operas mounted by Raoul Gunsbourg; and the visual arts, represented by the posters of Alfons Mucha and the works of Louis Tinayre. The prince always preferred the harshness of Nordic nature and the rigors of the great outdoors to the frivolous pleasures of the bourgeoisie and the cosseting climate of the Mediterranean. But he well understood the importance of pursuing his father's endeavors the while adding a little twist of his own.

A PRINCE FOR MODERN TIMES

The turn of the 20th century brought a new emphasis on physical activity that saw sports competitions take off as never before. The opportunity to expand on the idea of games was not lost on Albert I, an accomplished sportsman for whom sport was not sport without sportsmanship. Lampooned as a master-gambler in the Nice and Paris newspapers, his intention was to detach himself from the image of the casino by promoting the sporting events sponsored by the Société des Bains de Mer: the SBM or "society of sea baths" that owned and managed the Monte Carlo Casino. The SBM was indeed a long-term patron of sports, among them canoe racing, fencing, tennis, athletics, clay-pigeon shooting and also chess. In 1901 chessboards did battle on green baize when the casino hosted the acclaimed Monte Carlo chess tournament. The year 1911 saw the first edition of the Monte Carlo Rally, followed in February 1912 by the much-trumpeted Georges Carpentier versus Jim Sullivan European Championship title fight, Frenchman Carpentier and Englishman Sullivan trading punches for the entertainment of the emergent nobility and their commoner counterparts. Two other events that captured the prince's interest were the experiments conducted by Alberto Santos-Dumont in the skies over the principality; and the 1921 Women's Olympiad, the first international sporting event for women, held in Monaco on March 24-31. In the absence of a proper stadium or track, they took place in the clay-pigeon shooting ground and were contested by representatives from five countries: Great Britain, Switzerland, Italy, Norway and France. Under the enlightened leadership of French sportswoman Alice Milliat, the Games did much to debunk the stereotypes that gave women's sports a bad name. But the prince's ideals were not confined to the championing of sports, entertainment and sea bathing, though he was of course familiar with the health benefits of sea swimming. Albert I also wanted to lay the foundation for a modern Monaco; and while he was not politically inclined he had a firm grasp of the issues involved. Knowing that the first step was to reform and revamp Monaco's institutional system, he decided to grant the country a Constitutional Act making the principality a state under the rule of law. Though Albert I was less liberal in Monaco than in Paris and remained attached to the image of a hereditary, paternal ruler, his decision says a lot about his balanced outlook and complex thinking.

A son Altesse Sérénissime

Albert 1er

Prince de Monaco

Ouverture de Fête

l'inauguration du Musée Océanographique de Monaco

Ms. 662

C. Saint-Saëns

op. 133

C. SAINT-SAËNS

Ouverture de Fête

Ecrite pour l'inauguration du

Musée Océanographique de Monaco

OP. 133

	net		net
Partition d'Orchestre	15 »	Chaque Partie séparée	1.50
Parties d'Orchestre	25 »	2 Pianos a 4 Mains, par l'Auteur	

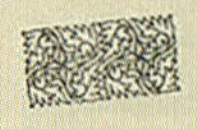

Paris, A. DURAND & FLS, Editeurs
DURAND & Cie
4, Place de la Madeleine

IMP. CHAIMBAUD - PARIS

LEFT-HAND PAGE
Score for *Ouverture de Fête*, written by Camille Saint-Saëns for the opening of the Monaco Oceanographic Museum, 1910. The composer's dedication reads "To His Serene Highness Prince Albert I of Monaco."

RIGHT
Prince Albert I of Monaco with Jules Massenet at the piano.

BELOW
Concert at the Monte Carlo Palais des Beaux-Arts, 1907.

Sports in the reign of Albert I

A sports enthusiast and lover of the great outdoors, the prince promoted sports development in the principality through his involvement with the Société des Bains de Mer.

LEFT-HAND PAGE
Top: Start of the 250 meters at the 1921 Women's Olympiad in Monte Carlo.

Bottom: Prince Albert I cycling in the Palace gardens, 1894.

RIGHT
From top to bottom:
A fencing match in Monaco, 1902; A game of golf; Tennis tournament in Monaco, 1902.

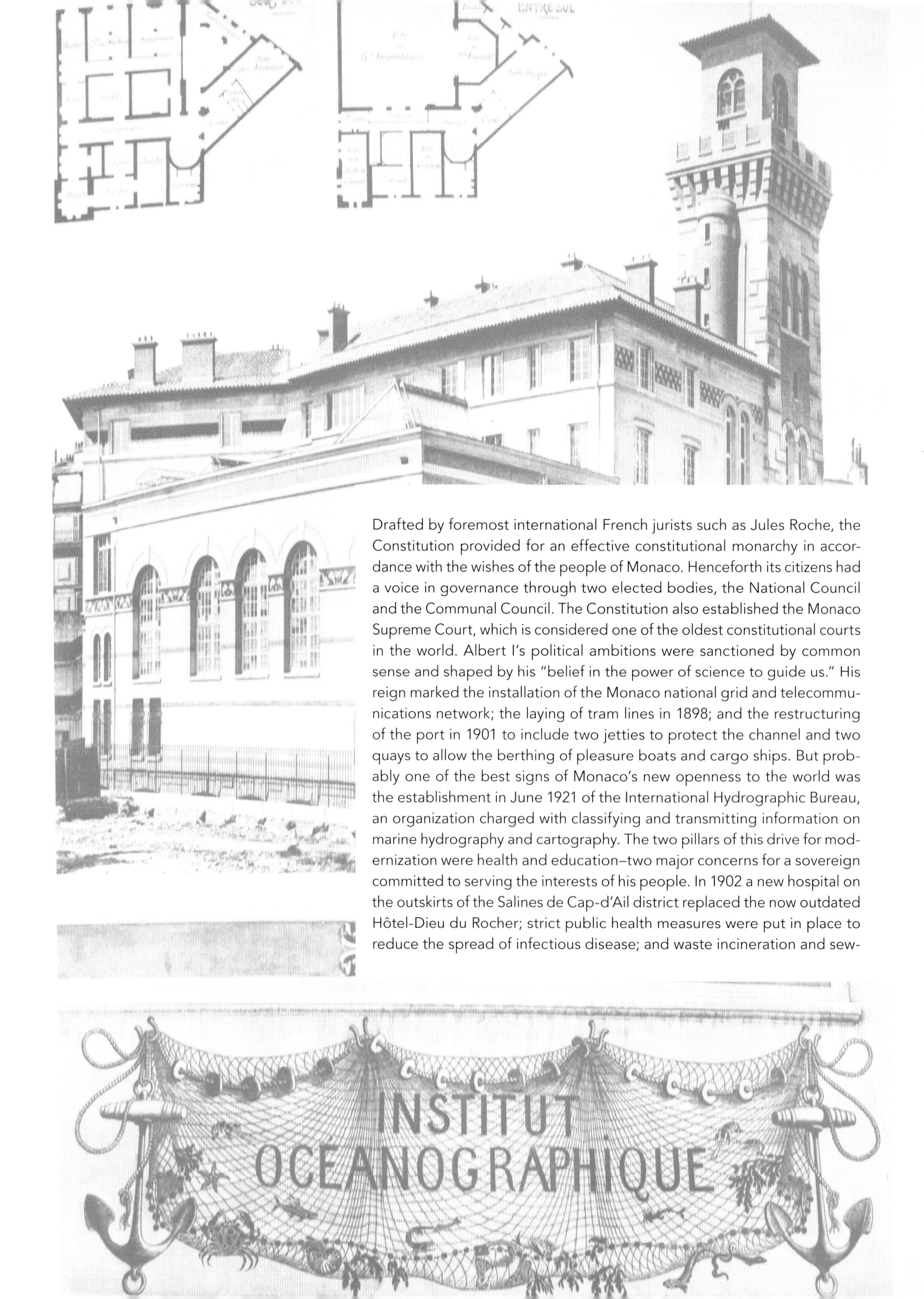

Drafted by foremost international French jurists such as Jules Roche, the Constitution provided for an effective constitutional monarchy in accordance with the wishes of the people of Monaco. Henceforth its citizens had a voice in governance through two elected bodies, the National Council and the Communal Council. The Constitution also established the Monaco Supreme Court, which is considered one of the oldest constitutional courts in the world. Albert I's political ambitions were sanctioned by common sense and shaped by his "belief in the power of science to guide us." His reign marked the installation of the Monaco national grid and telecommunications network; the laying of tram lines in 1898; and the restructuring of the port in 1901 to include two jetties to protect the channel and two quays to allow the berthing of pleasure boats and cargo ships. But probably one of the best signs of Monaco's new openness to the world was the establishment in June 1921 of the International Hydrographic Bureau, an organization charged with classifying and transmitting information on marine hydrography and cartography. The two pillars of this drive for modernization were health and education–two major concerns for a sovereign committed to serving the interests of his people. In 1902 a new hospital on the outskirts of the Salines de Cap-d'Ail district replaced the now outdated Hôtel-Dieu du Rocher; strict public health measures were put in place to reduce the spread of infectious disease; and waste incineration and sewerage systems were installed. In 1909 a library came into being, followed in 1910 by the Lycée Albert Premier: a prestigious public secondary school combining the advantages of the French and German education systems. The school was established on the recommendation of a report compiled by Gaston Moch at the request of the prince, who regarded education as the best defense against hatred and injustice. The teaching staff included former pupils of the École Normale Supérieure hand-picked by the prince to further the school's reputation. Among them was French writer and future laureate of the Prix Renaudot, Armand Lunel, appointed professor of philosophy at the Lycée. Yet another example of the prince's intuitive thinking was his recognition of the tourism potential of Monaco's new exhibition venues: the Museum of Prehistoric Anthropology, founded by Albert I in 1902; and the Oceanographic Museum, founded in 1898 and completed in 1910, described by the prince as "a vessel anchored to the shore filled with riches extracted from the depths." The direct counterparts of his Parisian institutions, these museums likewise served as showcases for the scientific works of the Sovereign Prince of Monaco.

LEFT-HAND PAGE
The Oceanographic Institute was founded by Albert I, Prince of Monaco on April 14 1906. The building on Rue Saint-Jacques was opened in the presence of President Fallières on January 23 1911.

BELOW
Laying of the first stone of the Oceanographic Museum, by the Count of Münster, representing Kaiser Wilhelm II. Construction commenced a year earlier. Monaco, April 25 1899.

The first Museum of Prehistoric Anthropology was located on the Rock of Monaco and run by Léonce de Villeneuve, a French canon, archaeologist and paleontologist who undertook archaeological excavations of the Grotte de l'Observatoire, in the heart of the Jardin Exotique. The findings attested to the presence of Paleolithic hunter-gatherers in the area now covered by the Principality. The Oceanographic Museum of Monaco meanwhile, under the leadership of the prince's friend and close associate Jules Richard, fulfilled a three-fold mission: to house the collections derived from scientific campaigns; to serve as a showcase for a new national identity; and to provide a venue for major conferences that testified to Monaco's international role. In April 1914 for instance, its elegant conference room hosted the first International Criminal Police Congress–the meeting that gave birth to the idea of Interpol. Then and now the museum's respective directors perpetuate the heritage of a prince committed to the preservation and protection of the origins of life.

Prince Albert I at work in his office, March 1905.

Enamored of the absolute and all things grand, the prince was no less attentive to detail. He well knew that appearances matter–that the allegorical embellishments of the façade of the oceanographic museum should be reflected through the prism of toponymy. To that end, in 1901 he named a street in the La Condamine district Rue des Açores–a name that placed the Azores so dear to the heart of this ocean-loving prince at the heart of the urban landscape. And on that note, let us now set off on the next stage of this adventure.

PORTUGAL and Norway

Destination passion

"We have already conquered the Sea,
It now remains only to conquer the Sky,
Leaving the earth to others …
Be as plural as the universe."

Fernando Pessoa, *Message*, 1934

n this late 19th century that saw the beauty of sail power reinforced by the power of steam, Monaco stood comparison with the most powerful nations in the world. While other European courts succumbed to the craze for yachting, as a sport but also to make a political statement, Prince Albert I nurtured an ambition that went far beyond pleasure alone. For him the sea was first and foremost a discipline. The rules of life at sea and the constraints imposed by the elements dampened his youthful ardor. By adding a third dimension, oceanography, he turned the sea from a simple resource into an object of study. After training as an officer in the Spanish Navy, the young crown prince cut his teeth as a sailor on his own boats, starting with *Isabelle II*, a small coaster designed for use as a training vessel. Then came four emblematic boats that were progressively equipped to meet the requirements of ambitious research projects: the *Hirondelle*, his first love, the boat he bought in Torquay in 1873 and on which he perfected his sailing skills before kitting it out for his first four scientific expeditions undertaken in the years 1885-1888; in 1891, the first *Princesse-Alice*, named after his second wife; the second *Princesse-Alice* in 1897; and the second *Hirondelle* in 1911.

PREVIOUS PAGES
Whaleboat in a bay
in Spitsbergen.

LEFT
The first *Hirondelle*,
a schooner acquired in 1873
and used by the prince
for his first four scientific
expeditions between 1885
and 1888.

After training as an officer in the Spanish Navy, the young crown prince cut his teeth as a sailor on his own boats, starting with *Isabelle II*, a small coaster designed for use as a training vessel. Then came four emblematic boats that were progressively equipped to meet the requirements of ambitious research projects.

Prince Albert I on the deck of the second *Princesse-Alice*. Photo by Charles Chusseau-Flaviens, 1905.

Tage Fl.
Elvas
Estremaz
Evora
Sétubal
Sines
Sadão R.
Béja
Guadiana
Ourique
Sierra de Monchique
ALGARVE
Lagos
Sagres
C. St. Vincent
Tavira
Faro
Huelva
Palos
C. Ste. Marie
PORTUGAL
ET
ESPAGNE
PAR L. VAT.

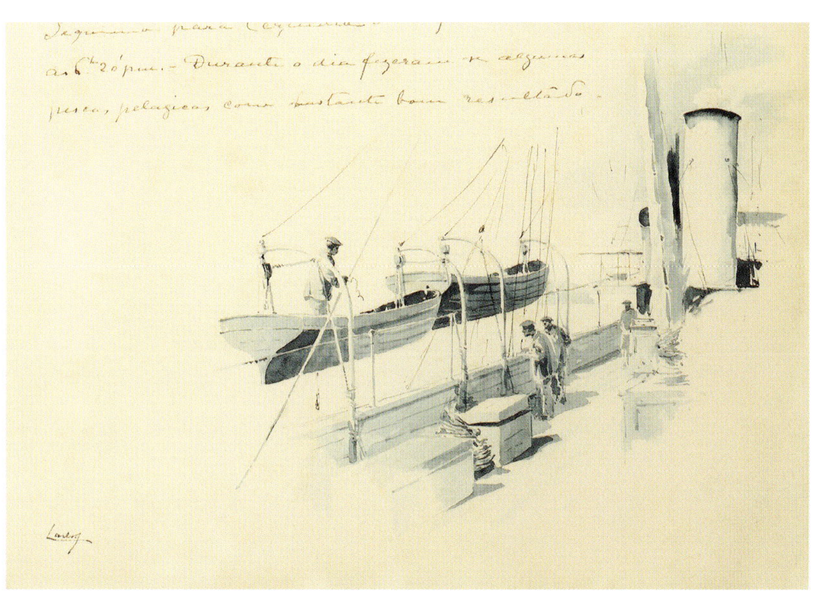

These last two boats, more powerful and propeller-driven to overcome the vagaries of the wind, enabled him to pursue his works with indomitable ambition and to assert himself as the "instigator and promulgator of oceanographic science." It was aboard these four boats that Prince Albert I led a series of 28 carefully documented expeditions detailing his observations, collections and experiments. Every expedition reflected his conviction that technology should serve to advance the cause of science, improve efficiency and produce new innovations. Planning the expeditions was no easy task, involving several planning stages, from determining the area of study to recruiting scientists with the skills required to meet the research objectives–which were different in every case because no two voyages were ever alike. Each one had its own particular flavor, aims and of course destination. What is more, they extended beyond exploration into the realms of politics, serving as a vehicle for representation and a display of power. Clad in full dress uniform, Albert I was seen as the embodiment of power at sea, not simply as a prince who was absent from his little nation. Expeditions were an opportunity to play to his strengths and capitalize on his capacity for action.

LEFT-HAND PAGE
"My life's work, much of it undertaken in Portuguese waters." (Letter from the prince to King Manuel, 1910).

ABOVE
Wash drawing by King Dom Carlos depicting the deck of his yacht *Amelia II*, drawn in spring 1897 to illustrate his logbook.

LEFT
The 1896 expedition aboard the first *Princesse-Alice*, off the coast of the Azores. In the foreground, Pierre Minelle and Jeanne Le Roux; in the background, from left to right, Henri Neuville, sailor Jean-François Le Quellec and Adolphe Fuhrmeister.

RIGHT-HAND PAGE
Top: Sailors on the bowsprit, at the prow of the second *Princesse-Alice*.

Bottom: Joint chiefs of staff gathered on the deck of the second *Hirondelle*, on August 4 1914, the day that France declared war on Germany. A few days later, the prince postponed this expedition and returned to Monaco. From left to right and top to bottom: Ranc, Bourée, Fuhrmeister, Tinayre, d'Arodes, Richard, Gain, de Rotschild, Louët; Taking time off for a photograph.

This dimension would take on particular importance in two regions which, for polar explorer Prince Albert I, could be considered as two parts of the same whole despite their markedly contrasting topography. It was on and under the seas off the coasts of Portugal and Norway that the prince pursued his vocation, meanwhile charting an original course that was entirely dictated by his abiding preoccupation with insularity. His was an initiatory journey to elsewhere, driven by the winds toward the Azores, Madeira and Spitsbergen, not as a modern-day Ulysses but as someone repeating ancient schemata in a bid to distance himself from the everyday and attain a mythical horizon–as evidenced by his account of island navigation, *La Carrière d'un navigateur.* It was his wish to shed new light on Macaronesia, the "Islands of the Blessed," and the Hyperborean lands, torn between a quest for an idealized nature and the scientific rigor of classical taxonomy. It was on those islands that the prince made important encounters. Madeira saw the dawning of his love for Alice Heine. The Azores marked the forging of a strong relationship with Captain Afonso Chaves. Lisbon brought him closer to King Carlos I–a friendship built on science. In northern regions, the same camaraderie flourished between the prince, his crew and his other companion travelers, among them scholars, the artist Louis Tinayre and Ship Captain Gunnar Isachsen.

PORTUGAL–A FIRST HORIZON

Prince Albert first came to Portugal in 1873, following the establishment of the first Consulate of Monaco in Lisbon by his father two years earlier. It was merely a pleasure trip, a chance to sail his boat the *Hirondelle*, but enough to ignite his curiosity about the region. He returned there a few years later, having spent the intervening time exploring the Mediterranean, then in 1879 he discovered Madeira. It was there, amid the cossetted luxury of a vacation spot beloved by the rich and famous, that he made the acquaintance of French politician Armand Chapelle de Jumilhac and his wife Alice Heine. Though the island itself found little favor in the prince's eyes, not so this young woman whose friendship he cultivated before eventually falling in love with her and making her his wife following the death of her husband in 1880. His next stop was Lisbon, which is probably where he first met Dom Carlos I, then 15 years his junior but who ascended the throne in the same year (1889). Here were two men united by a passion for oceanography who together and separately would produce a significant body of work. Both men were self-taught and free to go wherever they pleased, one aboard the *Princesse-Alice*, the other aboard the *Amelia*. Both men were equally careful to document their expeditions, comparing and contrasting their findings in a rich exchange of letters featuring illustrations by Dom Carlos. But Prince Albert, unlike his Portuguese counterpart, was blessed with a relatively trouble-free reign that allowed him to achieve far greater things.

Views of the archipelago of the Azores shot by Jules Richard.

From top to bottom: Ponta Delgada (São Miguel); Formigas Islets (to the south of São Miguel); Molho da baia, Ponta Delgada, June 1895.

"In March 1879, approaching Madeira in my schooner L'*Hirondelle*, I noticed a chain of little islands not far from the capital, Funchal. Their spiky profile stood out against the blue sky, looking like the dorsal fin of some monstrous fish beached on a sandbank."

La Carrière d'un navigateur, chapter V, "À la chasse", p.75, 1913 edition.

The prince's contribution to oceanography can mainly be explained by the length of his career at sea. In the 1880s, the majority of his expeditions and oceanographic campaigns took place in Atlantic maritime areas under the sovereignty or jurisdiction of Portugal: the Gorringe Bank, the Seine, Dacia and Josephine Seamounts and the archipelagos of Madeira, the Azores and Cape Verde. Between 1885 and 1914 he undertook no fewer than 14 campaigns in the vast archipelago of the Azores (Faial, Flores, Graciosa and São Miguel), transition zones between Europe and America much favored by naturalists as places to observe the dissemination of species. In 1888 and 1889 the main reason for his trip to Madeira was Alice's presence on the island; but as a scientist he took advantage of his stay to test new equipment he planned to use on his next expeditions. The years 1897-1912 saw him lead six campaigns that included the islands of Madeira, Deserta, Porto Santo and Selvagens. The expeditions were carefully coordinated by the prince and his teams, who would draft the work plan for the following year while the existing campaign was still underway, based on previous findings and reflecting the precision required for naturalistic studies, with particular attention paid to instrumentation. We could talk at length about the abundant research conducted by the prince, but this is not the place. Instead, we must content ourselves with looking at the two guiding themes of his expeditions: the study of the characteristics of the marine environment; and the distribution of organisms. The first step was observation: surveying, dredging and trawling; and capturing sea turtles and other animals using long lines, trammel nets and bar rigs.

ABOVE
Hauling up the stirrup-shaped trawl basket, a device for gathering deep-water samples employed on scientific expeditions from 1886 onward.

BELOW
Engraving depicting a *nasse triédrique*: a three-sided basket developed by the prince from 1888 onward and consisting of wood, wicker and netting.

RIGHT-HAND PAGE
Sperosoma Grimaldii Koehler: a form of deep-water sea urchin (echinoderm), discovered in the Azores in the course of the 1897 scientific expedition.

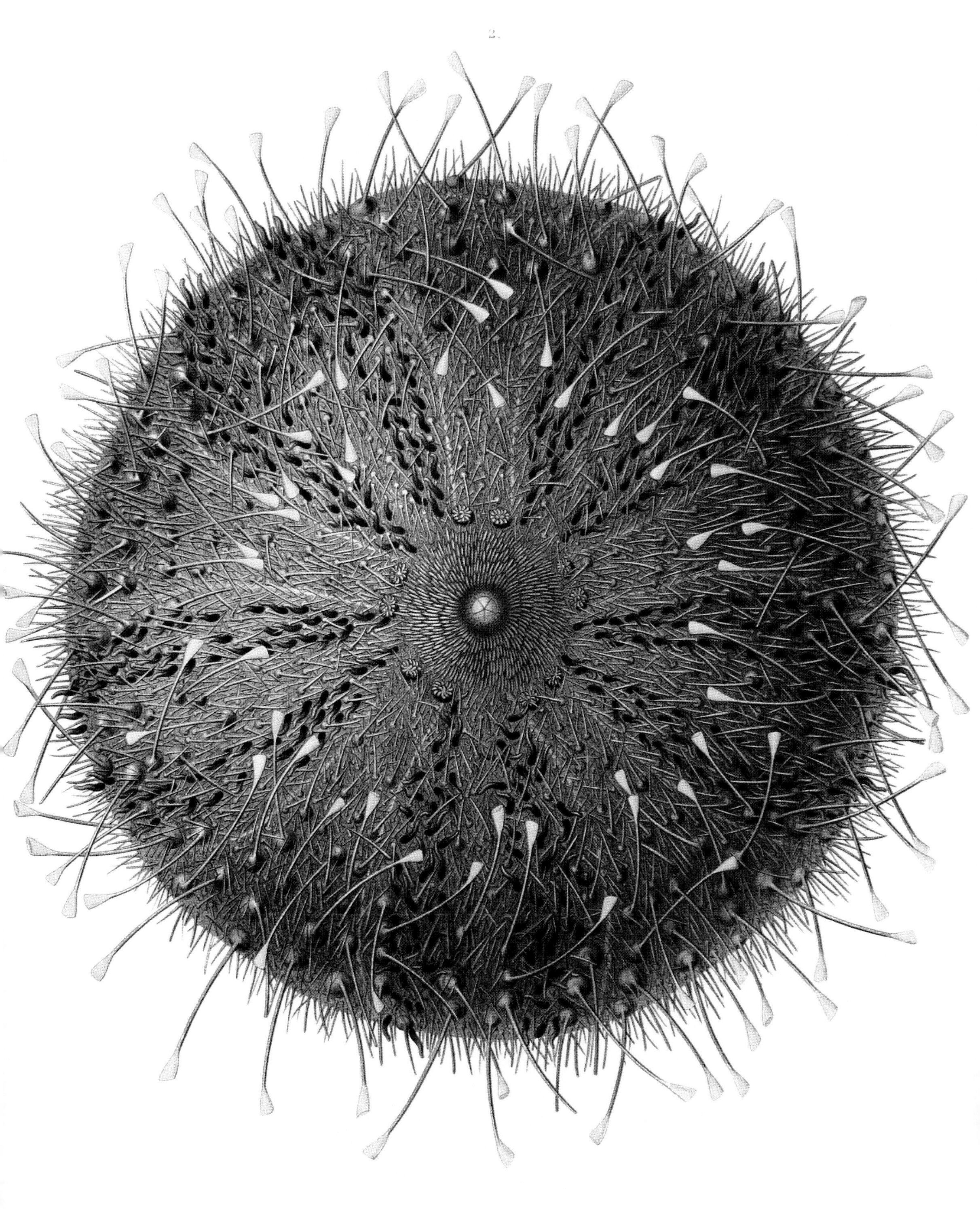

SPEROSOMA GRIMALDII KŒHLER.

La Carrière d'un navigateur, an account of the prince's travels in the Azores, features lavish descriptions of the creatures he observed, as here in the case of the moonfish: "having two, long fins, the ventral fin resembling a shipboard lifeboat while the dorsal fin looks exactly like a lateen sail, especially when it falls to one side as the fish is basking in the sun amid the rolling waves." Trading his compass for a pen, the explorer prince went into great detail about his art of observation. Then there was the topographical side of his expeditions: reconnoitering, naming and mapping areas in line with a systematic framework for exploring the region and lakes of the islands of the Azores. The prince's teams were as interested in insular areas and secluded enclaves as they were in the abyssal zones. They took measurements and collected samples at a depth of 5,500 meters in the Fosse de Monaco and conducted surveys to specify the area and relief of the Princess Alice Bank. Their temperature measurements revealed the existence of an oceanic trench exceeding 3,000 meters in depth near the Azores, between São Miguel and Terceira, which the prince named the Fosse de *l'Hirondelle*. Deep-sea trawling also yielded results, most notably on the day the expedition dropped its nets to the unprecedented depth of 6,035 meters and hauled up a fine collection of invertebrates plus an abyssal fish that was appropriately christened *Grimaldichtys profundissimus*. So it was that the prince made his contribution to the controversy that raged at the time between creationists and evolutionists over the potential for life in deep oceans. In the course of his observations he also provided convincing experimental evidence of new scientific phenomena, as

ABOVE
Left: "The *Hirondelle* under cloudy skies, at the dawning of oceanographic science." Bookstore publicity poster for *La Carrière d'un navigateur*, designed by Paul Berthon, Paris, Impr. Chaix, 1900.

Right: Illustration for *La Carrière d'un navigateur*, cover for the 1966 reprint, published by the Archives du Palais, with an introduction by Captain Cousteau, then director of the Oceanographic Museum (1957-1988).

RIGHT-HAND PAGE
French physiologist Charles Richet, 1850-1935.

in 1901 when Charles Richet and Paul Portier discovered the phenomenon of anaphylaxis. Sailing aboard the *Princesse-Alice* off the coast of Cape Verde, they observed that injecting low doses of *Physalia physalis* venom triggered a physiological response in which the immune system, whose job is to protect the body from infection, turned against its host causing the severe and sometimes fatal allergic reaction now known as anaphylactic shock. Their observation paved the way for a better understanding of the mechanism behind numerous disease processes and earned Richet the Nobel Prize in Physiology in 1913, thanks in part to the prince who funded the research.

"The Desertas Islands came into view – a view that unleashed a flood of memories!"

Prince Albert I, 1897

RIGHT-HAND PAGE
Top: Deserta Grande seen from the sea. Expedition of the second *Hirondelle*, 1912.

Bottom and following page: Prince Albert I during a hunt on Deserta Grande, between July 29 and August 1 1912.

— "As I discovered the innumerable life forms on this supposedly uninhabited island I had to wonder at the vanity of man for whom a deserted land is a land uninhabited by people [...] No, there is no such thing on Earth as a deserted land. The manifestations of life are everywhere to be seen, creating an infinity of riches that fills every corner of existence, from the deepest part of the ocean to the very outer limits of the atmosphere."

Albert I of Monaco, *La Carrière d'un navigateur*, chapter V, "À la chasse", p.106, 1913 edition.

SCIENTIFIC ADVANCES RESULTING FROM THE EXPEDITIONS

The prince strove to broaden the scope of oceanography, placing particular emphasis on the study of marine meteorology. In the course of his 1886 and 1887 expeditions he deployed several hundred buoys aimed at monitoring ocean currents, first on a line of latitude parallel to meridian 20 degrees west of Greenwich, then later between the Azores and the Grand Banks of Newfoundland. The results of his investigations were eagerly awaited and published in *La Nature* magazine (*Les recherches sur le Gulf-Stream. Visite aux Açores*, issue 676, year 14, 1886); his efforts had an impact in diplomatic and scientific circles alike. The prince deplored the lack of any coordinated international effort where marine meteorology was concerned and called for more accurate weather forecasts in the Azores–a region that henceforth served as a bridge between Europe and the North Atlantic thanks to the laying of the first telegraphic cables. As a keen supporter of telegraphy, the prince strove to implement a plan to create a meteorological observatory in the Azores, capitalizing on his friendship with Carlos I of Portugal and enlisting the support of Azores-based scientist, Captain Afonso Chaves. The observatory came into being in 1901, located in Horta on the island of Faial, and was subsequently named after the prince in 1923 following his death a year earlier. In 1904, the ocean's impact on upper-air weather was studied in the trade-wind belt; in 1905 weather balloons and pilot balloons were launched, together with weather kites carrying aluminum instruments designed to float high above the ocean (up to 4,500 meters). Quite apart from the rigor of the research itself, the sheer poetry of the undertaking would not have escaped the prince's attention–this man who describes himself in *La Carrière d'un navigateur* as being always on the lookout for "the future dawning on the ocean horizon, radiating a lofty ideal."

LEFT-HAND PAGE
Prince Albert I supervises the launch of weather or "sounding" balloons carrying instruments to take measurements of the upper atmosphere. First used in February 1905, this type of balloon would be employed throughout his subsequent expeditions.

ABOVE
Prince Albert reviewing material dredged up from the deep.

FOLLOWING PAGES
Cover and extract of *Résultats des Campagnes Scientifiques Accomplies sur son Yacht par Albert Ier, Prince Souverain de Monaco*, fascicule 38, 1912 (one of 110 fascicules first published in October 1889.)

RÉSULTATS

DES

CAMPAGNES SCIENTIFIQUES

ACCOMPLIES SUR SON YACHT

PAR

ALBERT IER

PRINCE SOUVERAIN DE MONACO

PUBLIÉS SOUS SA DIRECTION

AVEC LE CONCOURS DE

M. JULES RICHARD

Docteur ès-sciences, chargé des Travaux zoologiques à bord

FASCICULE XXXVII

Mollusques provenant des campagnes
de l'HIRONDELLE et de la PRINCESSE-ALICE dans les Mers du Nord

Par PH. DAUTZENBERG et H. FISCHER

AVEC ONZE PLANCHES

IMPRIMERIE DE MONACO

1912

R. Kœhler del. M. Borrel, Lovatelli Ch. B. de Monvel pinx. Werner & Winter, Francfort s. M.

1 PENTACRINUS WYVILLE THOMSONI 2 SOLASTER ENDECA VAR GLACIALIS
3 SOLASTER ENDECA 4 POROCIDARIS PURPURATA 5 PENTAGONASTER SEMILUNATUS
6 LINCKIA BOUVIERI

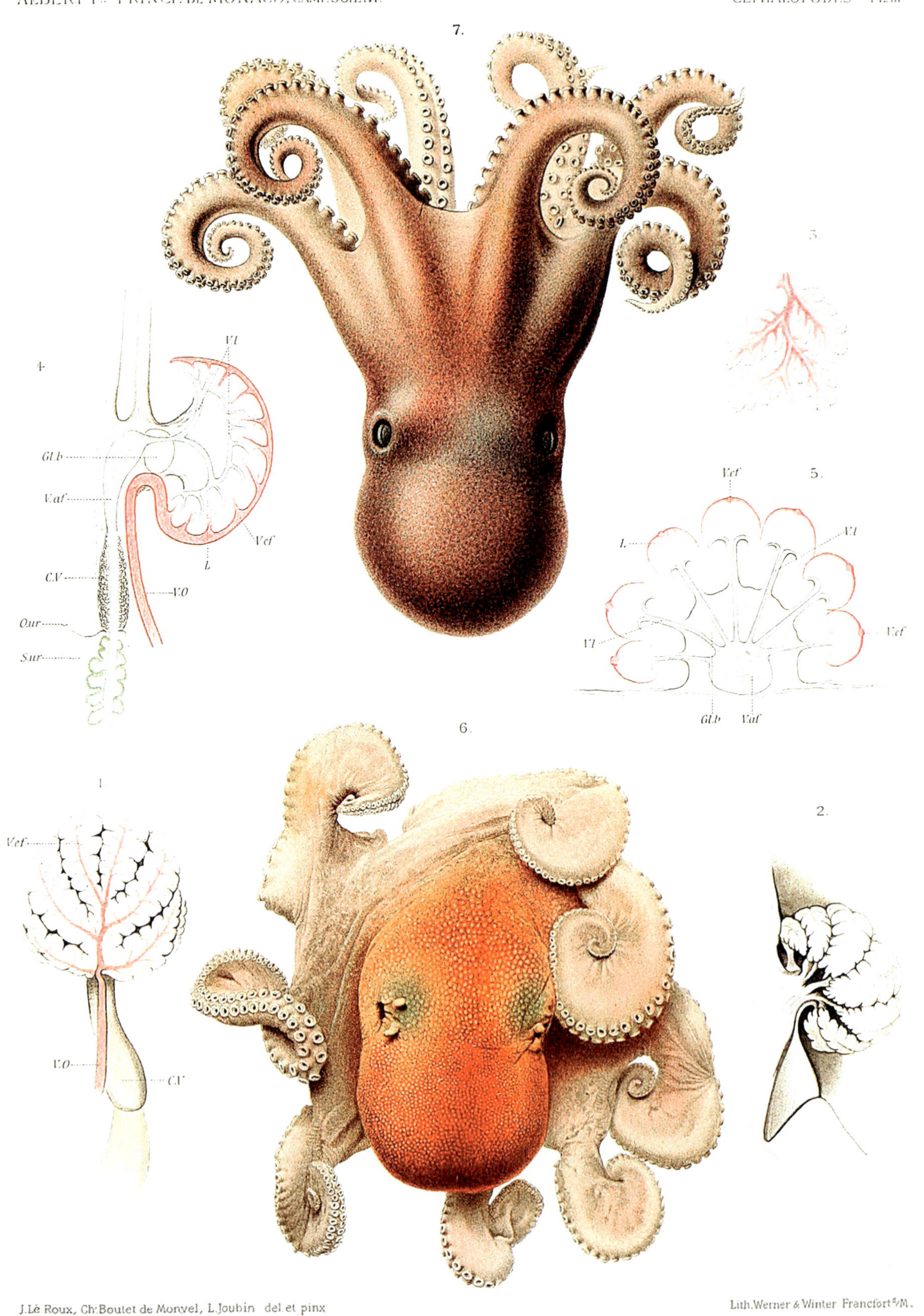

J. Le Roux, Ch. Boutet de Monvel, L. Joubin del. et pinx

Lith. Werner & Winter Francfort s/M.

1-5 CIRROTEUTHIS UMBELLATA P. FISCHER 6 SCŒURGUS TETRACIRRHUS (DELLE CHIAJE) TIBERI
7 OCTOPUS LEVIS HOYLE

Islands also served to anchor the prince, ground him in the world and provide a welcome respite from the hardships of life at sea. Hence his interest in the land itself: in the huge variety of crops and planting techniques; in the management of water resources that hovered between famine and feast; in the youthfulness of the terrain, ever at the mercy of volcanoes, earthquakes and the forces of Nature unleashed. He took pleasure in contemplating the scenery–a mixture of manicured and unspoiled lands and "strange" corners "in the middle of mountain forests" that he enthusiastically explored. He marveled at the generosity of Nature and the geological scars left by "capricious upheavals, flaming thunder, ash falls and the lava flows that created these islands." His observations echo the works of the geographers of his time, men like Konrad Malte-Brun and especially French geographer, writer and anarchist Élisée Reclus. Like the prince, Reclus combines geographical evidence with mythical elements, as in this passage from *Nouvelle Géographie universelle* (volume 12, page 4 of a 20-volume masterwork published across the period 1875-1894): "[…] if the hypothesis is correct, according to which eruptive mouths open along the fault lines formed by coastal processes, then the Azores, Madeira, the Canaries and Cabo Verde follow the contours of an ancient coastline–the boundary of the Atlantis geology." Prince Albert also considered geography from the perspective of ethnography, as one might expect of a man for whom scientific observation was largely inseparable from moral reflection. He placed particular emphasis on the description of island women: "Pictured here is one of these amiable creatures viewed from the front. Winter and summer alike, she is chastely wrapped in a long mantle of black cloth–an opaque barrier that no eye, no matter how piercing or how bold, is sharp enough to penetrate. Were it not for the face peeking out of the hood, even her nearest and dearest would not recognize her […]."

Page from *Résultats des Campagnes Scientifiques Accomplies sur son Yacht par Albert Ier, Prince Souverain de Monaco.*

Then there were his observations of island lifestyle, particularly the unusual modes of transport which, being curious by nature, he did of course try out. The *carros de cesto* of Madeira were a case in point: wicker toboggans pushed by two men, ideal for the winding roads in the mountains of Madeira. A sketch in the prince's diary, dated July 1905, perfectly captures the innocent pleasure he took in the experience, making him look like a character in a graphic novel. Keenly aware of the persistence of traditional values, the prince did not fail to notice that education was essentially reserved for the privileged classes–an issue he raised with Portuguese revolutionaries following the fall of the monarchy in 1910. Despite his sadness at the assassination of his friend Carlos I of Portugal, the prince remained strongly attached to the country and was anxious to remain on good terms with its new Republican leaders. Ten years later indeed he would stay in Lisbon for three days (November 5-8) as the official guest of the then President of the Portuguese Republic, António José de Almeida–a fitting punctuation to his final oceanographic campaign, sailing aboard the *Giralda*, an *aviso* (dispatch boat) placed at his disposal for the duration of his investigations by Alfonso XIII King of Spain.

LEFT-HAND PAGE
Landing stage in Ponta Delgada, São Miguel, June 1895. Photograph by Jules Richard.

RIGHT
Funchal, Madeira, August 4 1912. Following luncheon at the Blandy family's Quinta do Palheiro estate, the prince and his party are pushed downhill in *carros de cesto*.

PAGES 84-85
Prince Albert I on one of his expeditions to Portugal.

PAGES 86-87
Prince Albert I on an official visit to Lisbon. Oil on canvas by Louis Tinayre, 1920.

— "Thrilled at the prospect of finally fulfilling my dream, I felt as though my life were suspended between a past life filled with the struggles, affections and pain of reality on terra firma, and a future life just dawning on the ocean horizon, illuminated by a lofty goal."

La Carrière d'un navigateur, chapter VI, "Le dernier voyage scientifique de l'*Hirondelle I*", p.124, 1913 edition.

E.FLORIAN
LOUIS TINAYRE

NORWAY: THE ARCTIC ADVENTURE

Norway was the second stage in the prince's Atlantic adventure, or rather the continuation of a quest for an *alma mater* ("nourishing mother") that saw him venture beyond the Azores and Portugal, even if he did eventually return there. By 1897 he already had his eye on more remote destinations, as evidenced by the letter he wrote that year to Chaves: "I have just returned from a trip to England where I attended the launch of my new boat, which must be ready in time for April; whilst there I took the opportunity to consider the requirements for an expedition in Arctic waters [...]." But visiting Norway was certainly uppermost in his intentions, a reflection perhaps of the then current fascination with Scandinavia as a locus for Norwegian romantic nationalism–a movement that saw the Viking saga as a new mythology for the present. The prince's fascination for the polar regions is exquisitely expressed in the final chapter of *La Carrière d'un navigateur*, where he makes use of anaphora to convey the full force of his feelings: "I love the North for its power to attract men far away from unjust acts [...] I love the North for its limpid light so soothing to the eyes [...] I love the North for its dignified silence undisturbed by the passage of death [...]."

The Arctic embodied a long-cherished dream for Prince Albert I, one nourished by the books he bought from the bookshop in Orléans near the Collège La Chapelle-Saint-Mesmin, particularly the biography of Alexander Mackenzie, and Charles Francis-Hall's *Life Among the Esquimaux* recounting his expedition in search of Sir John Franklin. It would be a while, however, before he could achieve his dream. First came a string of disappointments, most notably his unsuccessful attempts to join Adolf Nordenskiöld's Arctic expeditions in the 1870s. Undeterred, the prince honed his navigational skills, equipped a ship for Arctic conditions and in 1898 his childhood dream finally came to fruition. That year saw the mounting of the first scientific expedition to Spitsbergen (the largest island of the Svalbard archipelago) all the way to Moffen Island off the northwest coast of Spitsbergen. Three further expeditions were to follow, in 1899, 1906 and 1907. All of them were conducted aboard the second *Princesse-Alice*, a powerful ship launched on November 27 1897 at the Laird shipyard, in Birkenhead near Liverpool, England. With a top speed of 13 knots, the ship combined a steel structure 73.15 meters long with a 100 horsepower triple expansion steam engine, a propeller and two masts (three yards on the main mast). According to the ship's logbook, the crew henceforth included mechanics alongside sailors, cooks and laundry staff.

"On many occasions my ship had to sail across ice-covered waters ..." Engraved illustration by Ernest Florian from a drawing by Louis Tinayre, p.282, chapter 8, *La Carrière d'un navigateur*; 1913 edition.

OCÉAN GLACIAL ARCTIQUE
SPITSBERG (CÔTE NORD-OUEST
Carte dressée sous la Direction de
S.A.S. le PRINCE DE MONACO
pendant les saisons d'été 1906 et 1907
par
la Mission ISACHSEN
composée de
M.M. GUNNAR ISACHSEN, Chef de la Mission, Capitaine de cavalerie de l'armée Norvég
ARVE STAXRUD, Lieutenant d'infanterie d°
FERDINAND LOUET, Médecin Major de l'armée Française
HANS HORNEMAN, Géologue.
ADOLF HOEL, Géologue.
ALFRED KOLLER, Ingénieur.
ALV STRENGEHAGEN, KARL HAAVIMB, ANDERS LOSVIK.
HAAKON MYHRE.
Sondages côtiers par M. BOURÉE Lieutenant de Vaisseau
LÉGENDE
I. D'AMSTERDAM
I. DES DANOIS
BAIE MAGDALENA
Presqu'île des Rennes
Iles des Canards
Ile de Station
Iles des Mouettes
Iles Lerner
LIEFDE
BAIE BOCK
Plateau
Monts du Président Loubet
BAIE KING
Iles Lovén
DÉTROIT DU FORELAND
(DÉTROIT FOUL)

LEFT-HAND PAGE
Map of Spitsbergen (northwest coast), drawn up on 1906 and 1907 expeditions, awarded first prize in the 1910 Bergen tourist and sports exhibition. *Résultats des Campagnes Scientifiques*, fascicule 40, 1912.

ABOVE
Gunnar Isachsen, oil on canvas by Louis Tinayre.

For Prince Albert I, Spitsbergen was the ultimate expedition, embodying the very essence of his emotional geography. It is this image of Prince Albert that survives today, an image forever associated with the heroic age of polar exploration when men like James Cook, Robert Edwin Peary, Roald Amundsen and Fridtjof Nansen vied for supremacy in the Polar Regions. Prince Albert I was truly himself in Norway, a place where he could distance himself from the Mediterranean that he disliked. Setting aside scientific research, Nordic countries provided an opportunity to test himself, rise to the challenge of the harsh conditions and discover who he really was. Find his way to find out more, know himself the better to follow the laws of Nature. Obstacles, fog, glaciers, cold temperatures and boulders piqued his curiosity. And these were special times for the prince. At the turn of the 20th century he was a seasoned fifty-year-old with a wealth of experience behind him; his curiosity and versatility knew no bounds. It was in Spitsbergen, despite its remoteness from European capitals, that he learned of the notorious Dreyfus Affair. Every stopover found him scouring the international newspapers and correspondence to keep abreast of developments. Eager to play his part, in 1889, shortly before the start of the trial in Rennes, he wrote a letter to Flore Singer expressing his concern for Alfred Dreyfus, followed on June 27 that year by a letter addressed to Lucie Dreyfus, sent from Kiel, which he then had published in *Le Figaro* on July 3. Some years later, in 1906 in Adventfjorden (Advent Bay), he learned of the captain's "rehabilitation." He meanwhile governed from a distance, the principality ever uppermost in his mind no matter how far he traveled. He also continued to indulge his passion for hunting, though as evidenced by this letter from Princess Alice, his youthful ardor was now tempered by increasing concern for wildlife: "It certainly sounds like a wonderful journey, if a bit cold and bleak with all that ice, but then you are doing what you love, doing what you have always said you wanted to do. I thought you were going to treat yourself to killing a few bears. Thank God you have given up on that idea! For someone like you, traveling to the Far North must be a particular thrill, more so than for anyone else and I am happy for you." (*Archives du Palais de Monaco* [APM] C 650).

RIGHT-HAND PAGE
Top: "The snow was granular." Engraved illustration by Ernest Florian from a drawing by Louis Tinayre, p.317, chapter 8, *La Carrière d'un navigateur*; 1913 edition.

Bottom: "Four very poor Norwegians came to settle in a cabin." Engraved illustration by Ernest Florian from a drawing by Louis Tinayre, p.303, chapter 8, *La Carrière d'un navigateur*; 1913 edition.

LOUIS TINAYRE
EFLORIAN

LOUIS TINAYRE
EFLORIAN

1.

Fridtjof Nansen. Godthaab, Lysaker 2 July 1898.

Albert I
Sovereign Prince of Monaco.

Dear Prince Albert,

Please accept my most hearty thanks for your honouring letter and extremely kind offer to help me in my work by making some special observations in the North Ocean. It was a great delight to me to learn that Your Highness are going North on an Arctic cruise with the "Princesse Alice" this year, as I know that we will then get some highly important and much needed results observations from this part of the ocean, some of which I have been specially looking forward for as they are necessary to be able to explain some of conditions found in the North Polar Basin during the drift of the "Fram".

As I know that your time is precious I shall only draw your attention to some peculiarities in the temperature of the sea in the various depths, which I think it would be of great importance to have carefully reexamined.

Your Highness will probably know, that we found comparatively warm water under the cold layer of surface water of the Polar Basin. The section on the enclosed Pl. I will probably give an impression of the average distribution of the temperatures from the surface to the bottom as we found them in the Polar Basin. Between ~~100m &~~ 200m & and 500m the temperatures were from +0.5°C to +1.15°C. Under 500m the temperature gradually and slowly sank with the depth. At 900m it was 0.0°C. At 2000m about ÷0.67° to 0.70°C. At 3000m between ÷ −0.73 and −0.82°C. At 3800m −0.64°C.

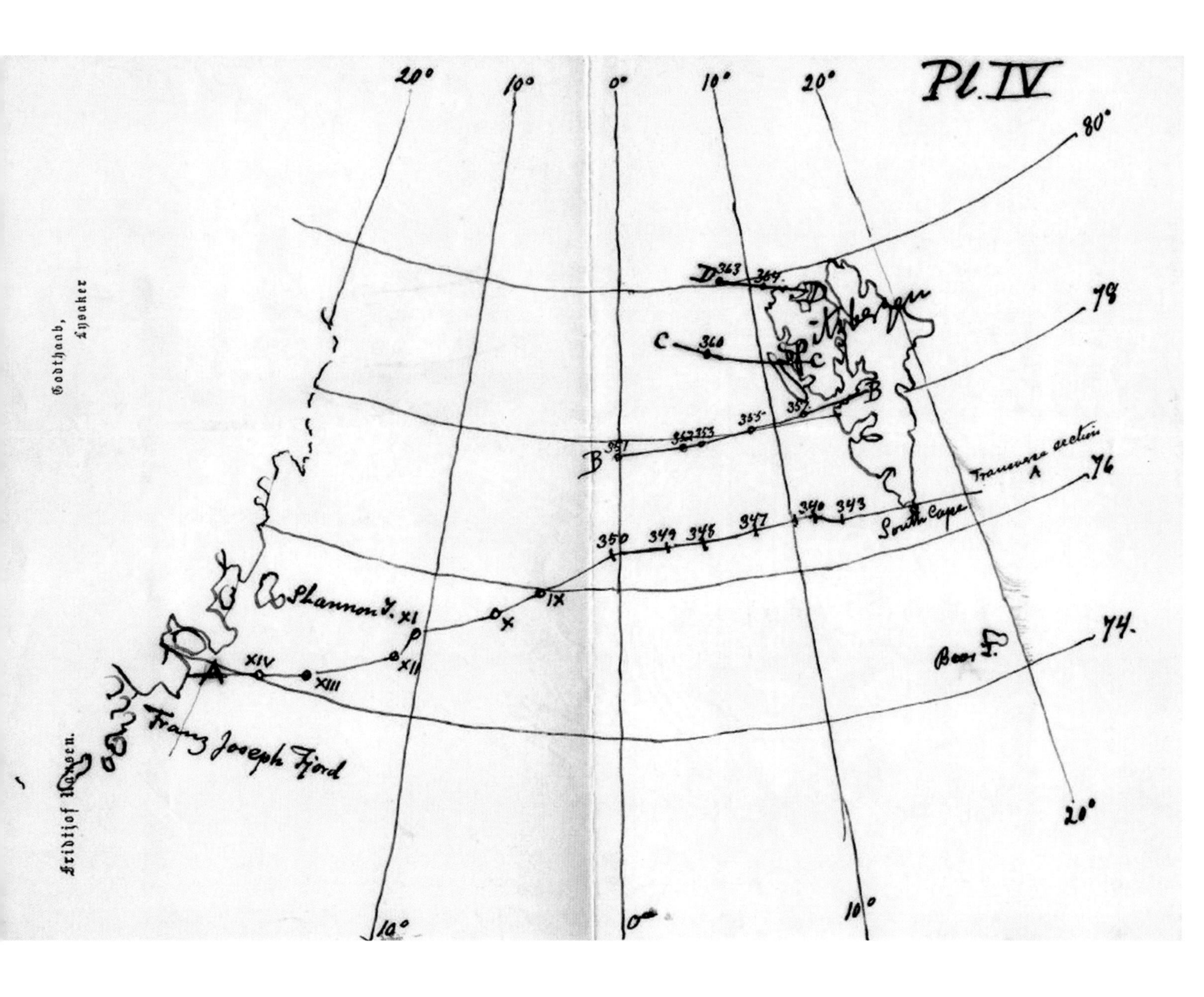

Letter from F. Nansen dated July 2 1898 in reply to the prince's letter of June 22. The letter is accompanied by a sketch and urges the prince to take measurements using the new, high-performance instruments on board the second *Princesse-Alice*, in order to compare the results with Nansen's own findings.

ABOVE
Photo shoot. Cross Bay, August 24 1906.
Photo by Jules Richard.

RIGHT-HAND PAGE
Itinerary followed by the first and second *Princesse-Alice* in the period 1891-1899, sailing across the North Atlantic, the Mediterranean and the icy waters of the Arctic Ocean.

Then of course, there was his ongoing commitment to scientific research. Though he was not the first man to set foot in this *terra incognita*, Prince Albert I was no less a pioneer, particularly in the field of cartography and the etymology of fjords and "pointed mountains"–the literal translation of "Spitsbergen". The success of his four expeditions is a testament to the meticulous planning that went into them, from the careful selection of his teammates and equipment, to the detailed analysis of the findings recorded by his predecessors. He may well have taken inspiration from his first cousin, Wilhelm Karl, Duke of Urach, who in 1891 took part in Wilhelm Bade's expedition to Bear Island and Spitsbergen. Consulting specialists in the Polar Regions was a priority for the prince and his teammates. Jules Richard, for instance, one of the prince's most faithful companions, lost no time in seeking advice from Charles Rabot, a French geographer who had traveled to Spitsbergen on the *Petit Paris* in 1882, then again in 1892 aboard *La Manche* under the command of Amédée Bienaimé. The prince himself penned a number of letters to experts, among them Sigurd Scott Hansen, the Norwegian naval officer who took part in Fridtjof Nansen's expedition aboard the *Fram*. On June 22 1898, as evidenced by a letter now held at the University of Oslo Library, the prince sought the advice of Nansen himself, eager to hear the recommendations of the man who had

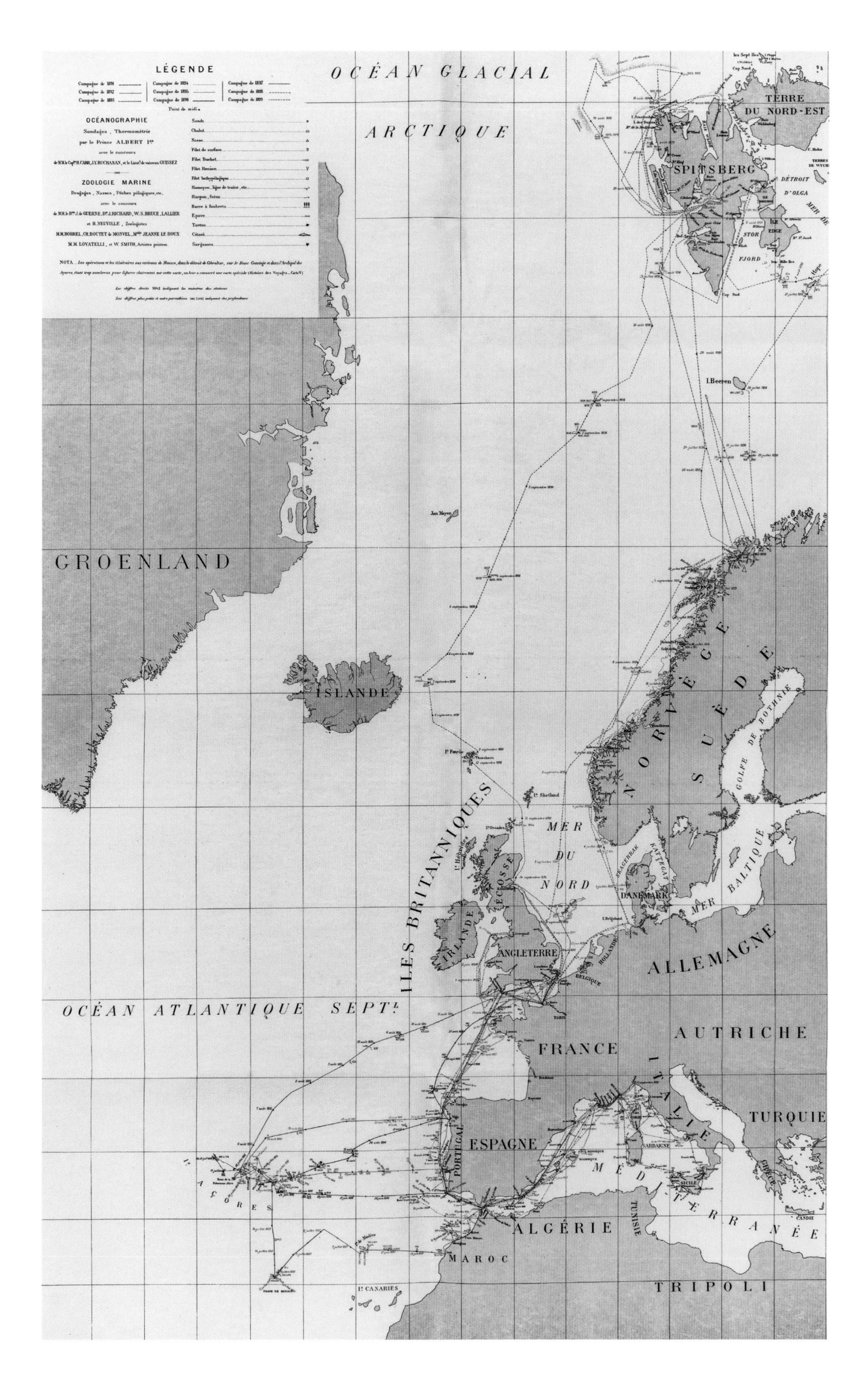

LÉGENDE
OCÉANOGRAPHIE
Sondages, Thermométrie
par le Prince ALBERT Ier
ZOOLOGIE MARINE
OCÉAN GLACIAL
ARCTIQUE
TERRE DU NORD-EST
SPITSBERG
DÉTROIT D'OLGA
STOR FJORD
I. Beeren
Jan Mayen
GROENLAND
ISLANDE
NORVÈGE
SUÈDE
GOLFE DE BOTHNIE
ILES BRITANNIQUES
MER DU NORD
ÉCOSSE
IRLANDE
ANGLETERRE
DANEMARK
MER BALTIQUE
ALLEMAGNE
BELGIQUE
HOLLANDE
OCÉAN ATLANTIQUE SEPTl
FRANCE
AUTRICHE
ITALIE
TURQUIE
ESPAGNE
PORTUGAL
AÇORES
MÉDITERRANÉE
ALGÉRIE
TUNISIE
MAROC
Ies CANARIES
TRIPOLI
CANDIE
SICILE
SARDAIGNE

established himself as the accepted authority on polar exploration following his expedition to the North Pole in the period 1893-1896. Nansen's reply (July 2 1898, Archives of the Monaco Oceanographic Museum) contained a sketch and urged the prince to take measurements using the new, high-performance instruments aboard the *Princesse-Alice* in order to compare the results with Nansen's own findings.

Prince Albert's maiden expedition, in 1898, was devoted to building up his collections in view of the forthcoming opening of the Monaco Oceanographic Museum (construction work began the following year). It consisted in the collection of Arctic, marine, terrestrial and freshwater animals, and in conducting observations on the glaciers relating to the formation of icebergs. It also provided the inspiration for the final chapter of *La Carrière d'un navigateur*–an opportunity for the prince to conduct scientific research but also to experience the metaphysical wonders of a journey that took him from Bear Island to Adventfjorden passing by way of Barents Island and Smeerenburg Bay. Far removed from the passions of men, he delighted in the silent serenity of the majestic landscapes and the evocative power of the place names. The Arctic spoke to his imagination as a writer, just as it had inspired Jules Verne (*The Adventures of Captain Hatteras*, 1866) and, chiming with his own firsthand experience, Léonie d'Aunet (*Voyage d'une femme au Spitzberg*, 1854). However, the scientist within him took pains to analyze and compare the results of the expedition with those of his Atlantic campaigns. As he wrote in his letter to Carlos I of Portugal: "My findings showed that many of the forms I had gathered in the Atlantic at depths of 1,500 to 2,500 meters were present in Arctic Waters along the shoreline. They also showed much stronger development."

The prince and his joint chiefs of staff on board the second *Princesse-Alice*, somewhere off the coast of Norway, in the period July 3-6 1906. On the left of the prince is William S. Bruce, leader of the Scottish expedition; on his right is Gunnar Isachsen, leader of the Norwegian expedition. Photograph by Henry Bourée.

"And one of the icebergs floated right past us." Preliminary drawing by Louis Tinayre for the 1913 edition (featured in an engraving by Clément on p. 275 of *La Carrière d'un navigateur*).

"We forged ahead through the ice field." Preliminary drawing by Louis Tinayre for the 1913 edition (featured in an engraving by Boileau on p. 293 of *La Carrière d'un navigateur*).

(p 346)
Nous pénétrâmes dans le champs de glaces,
et bientôt il fallut manœuvrer de cent façons
pour contourner les blocs petits et grands

ALBERT Ier PRINCE DE MONACO. CAMP. SCIENT. EXPLOR. SPITSBERG_TOP. GÉOL. PL. XIV

Isachsen phot.

1_Baie Cross (le 28 Juillet 1907)

Isachsen phot. Héliog. L. Schutzenberger

2_Baie Hamburger (le 18 Août 1907)

ALBERT IER PRINCE DE MONACO, CAMP. SCIENT. EXPLOR. SPITSBERG._TOP.GÉOL.PL.XV

Izachsen phot.

1 _ S^et Hoel (le 14 Août 1907)

Izachsen phot. Héliog. L. Schutzenberger Paris

2 _ Vallée Signe (le 13 Août 1907)

Panoramic views of northwest Spitsbergen.

The second Spitsbergen campaign, in 1899, was more technical and focused on physiological and bacteriological research, topographic and bathymetric surveys in Red Bay, and the discovery of Richard Lake. Noting the inadequacies in existing maps, it was the prince's wish to conduct a combined topographic and bathymetric survey of one of the fjords in Spitsbergen. To that end he retained the services of ship-of-the-line lieutenant Théodore Guissez, a specialist in his field who promised to deliver results that would be of practical help to future navigators.

The third expedition, in 1906, studied the upper atmosphere of the Arctic and featured a topographic and bathymetric reconnaissance of Cross Bay. Sailing with the prince was Louis Tinayre, a painter and illustrator he had met at the 1900 Paris World Fair and recruited four years later. Tinayre's sketches of the expedition would later feature in an illustrated edition of *La Carrière d'un navigateur*. The prince also took along Gaumont's new motion picture camera to capture footage of the Smeerenburg Glacier. He meanwhile sponsored two further Spitsbergen reconnaissance expeditions, one Scottish, led by William Speirs Bruce, the other Norwegian, led by Captain Gunnar Isachsen. On the return journey he wrote to King Haakon of Norway congratulating Isachsen on his achievements and results. A man of great political acumen, the prince was keenly aware of the importance and symbolism attached to such a success by a country that had only regained its independence the previous year.

ABOVE
Doctor Richard on the bridge of the second *Princesse-Alice* in 1899.

RIGHT-HAND PAGE
Left: Captain Carr and William S. Bruce on board the second *Princesse-Alice* in 1906.

Right: Preparing to take soundings.

Bottom: Spitsbergen: the dinghy being towed by "la pétroleuse", 1898.

The prince's fourth campaign in 1907 pursued and broadened the scope of the 1906 investigations to include numerous observations of changes in surface temperature as sea ice expands. The ambitious Norwegian expedition meanwhile entered its second year, Isachsen's team now reinforced by two new recruits: geologist Adolf Hoel; and botanist and photographer Hanna Resvoll-Holmsen, whose study of the Svalbard flora revealed the presence of rare species in Cross Bay despite its extreme northern location and predominantly rocky terrain.

Health problems and advancing age would prevent the prince from undertaking further expeditions in the Spitsbergen region, though the final leg of his 1908 campaign did take him all the way to Trondheim. Four years later, sailing aboard the second *Hirondelle* with his granddaughter,

ABOVE
Botanist and photographer Hanna Resvoll-Dieset in Magdalena Bay in 1907. Considered Norway's foremost ecologist, she accompanied the prince's 1907 expedition.

RIGHT-HAND PAGE
"An iceberg", preliminary drawing by Louis Tinayre, engraving by Boileau on p. 341 of *La Carrière d'un navigateur*, 1913 edition.

— "I love battling against the forces of the wind-whipped waves cleansed by snow that leave the soul prouder and more generous. I love the North because Death passes by with the dignity of silence, tenderly interring in fields of crystal those ravaged by the lies of the world."

La Carrière d'un navigateur, chapter VIII, "Croisière dans les régions arctiques", p.264-265, 1913 edition.

Mademoiselle Charlotte de Valentinois, he cruised along the Norwegian coast for the last time. And with that, his Scandinavian adventure came to an end. Not so his interest in the region, where he remained present in mind if not in body. In 1910, for instance, his expedition maps won first prize in the Bergen tourist and sports exhibition. He also continued to provide funds in support of exploration, committed to redefining the world and leaving his mark on the map of time. Nodding to that ambition are the place names derived from his expeditions, either in his lifetime or after his death: the Glacier Louis-Tinayre, the Glacier Loüet, the Glacier de Monaco, the Pic Prince-Albert, the Monts de l'Empereur-Guillaume and the Monts du Président-Loubet. While some of these places are now under threat from climate change, there is no better proof that Prince Albert I deserves our undying gratitude for his invaluable scientific and political contribution. They also serve to remind us of the urgent need to protect those parts of the world scarred by mining and undergoing accelerated change due to their particular location. This call has plainly reached the ears of the international scientific community, which has proved especially influential in supporting the establishment of the "Global Seed Vault" in the mountain above Longyearbyen. Holding more than one million samples of seeds, the vault ensures the genetic diversity of our food crops.

ABOVE
Recherche Bay, Spitsbergen, 1899. Joint chiefs of staff of the second *Princesse-Alice* standing on an ice floe.

RIGHT-HAND PAGE
Vue du Spitzberg by Louis Tinayre, 1912.

Other WORLDS

Lands and seas forever revisited

"He takes the mystic ocean at his feet
for the visible image of that deep, blue,
bottomless soul, pervading mankind and Nature."

Herman Melville, *Moby Dick*, 1951

In this last chapter, we focus on the prince's other destinations, presented not as a tedious catalog, more as an entertaining list of journeys that he undertook for various reasons, helped by his solid language skills. Some were for pleasure (hunting trips, attendance at the Congress of French Learned Societies); some were for the purpose of political representation; some were to satisfy his curiosity as a geographer. All of his journeys, however, all of his actions and all of his writings reflect his belief in Hugo's principle that what matters is "what you see." For the prince, the mark of a true traveler was precisely this readiness to stop and observe or, when circumstances demanded it, make a detour to rescue a fellow seafarer. As evidenced in his writings, he deplored hurried expeditions that took no account of the places and people in their path. For example, he recalls with great sadness crossing the Straits of Gibraltar and watching the Cape Spartel lighthouse recede into the distance. We will therefore confine ourselves here to those places of particular interest to Prince Albert I: the places where he lingered, not just places along the way.

PREVIOUS PAGES
The Erie Canal in Syracuse, New York State, 1900.

RIGHT-HAND PAGE
Engraving representing the Cap Spartel lighthouse built in 1864.

SPAIN: WHERE THE MEDITERRANEAN MEETS THE ATLANTIC

Spain occupied a special place in the prince's travel itinerary. He first went there in 1862 to visit Madrid and Toledo, accompanied by the future first Bishop of Monaco, Abbé Charles Theuret. Four years later he joined the Spanish Royal Navy as an ensign and spent the next two years sailing from Cadiz to the Caribbean and back under the expert guidance of navy lieutenant Simón de Manzanos. He quickly rose through the ranks, promoted first to ensign then to lieutenant and ultimately, in 1912, to the distinguished rank of rear admiral. Eight years later it was in full dress uniform that he made his final voyage to Lisbon to attend the gala evening organized under the auspices of Portugal's Ministry of the Interior. He returned to Madrid in January 1878 to represent his father at the marriage of King Alfonso XII, this time sporting the *Cruces del Mérito Naval*: a Spanish military award for merit in time of peace that celebrated the strong links between the principality and the Spanish monarchy. There are letters attesting to the long correspondence between the prince and Alfonso XII and especially his son, Alfonso XIII. The young Prince Albert had watched with dismay as Spain's ancient monarchy had reeled under threat of revolution on one side and Carlist extremism on the other–lessons that no doubt steered him toward modernity and balance when developing his own political doctrine. Following his investiture as sovereign prince of Monaco, he would form close ties with committed Republican, Spanish oceanographer, Odon de Buen, putting aside all political considerations for the sake of friendship and the pursuit of knowledge. In 1904 the Cantabrian

ABOVE
Abbé Breuil: the "Pope" of Prehistory.

RIGHT-HAND PAGE
The prince in the famous art-filled cave of Cueva del Castillo, Puente Viesgo, Cantabria, July 23 1909. With the prince, the prehistorians Henri Breuil and Hugo Obermaier as well as Hermilio Alcalde del Río, the archaeologist who first discovered the cave.

Pyrenees also provided an outlet for his scientific creativity, science being a quest for truth that is largely inseparable from politics. That year the prince lent his support to the work being conducted by Abbé Breuil and Émile Cartailhac on the prehistoric paintings and parietal art of the Altamira Cave in Santillana del Mar, Cantabria. Five years later he went there himself, stopping first in Puente Viesgo to visit the Cueva del Castillo then continuing to Santillana del Mar where he marveled at the decoration of what is known as the "Sistine Chapel of Quaternary Art." The experience reinforced his wish to promote prehistoric research, with the Institute of Human Paleontology opening a year later. The year 1912 found him once again in Madrid, this time as a guest of the Royal Geographical Society. In his element here, the prince renewed his commitment to the scientific exploration of the Mediterranean: "I now seek the agreement of all Mediterranean nations to participate in the oceanographic study of the Mediterranean, in line with the wishes of the International Geographic Congress that has entrusted me with the presidency of two committees charged with bringing to fruition this landmark project to study the Atlantic Ocean and the Mediterranean alike." After the collapse of the initial talks held in Rome in 1914, it was also in Madrid, at a conference held there on November 17-20, that the Commission Internationale pour l'Exploration Scientifique de la Méditerranée (Mediterranean science commission or CIESM) finally came into being.

MOROCCO AND STOPS IN TANGIER

From Spain we quite naturally head straight for the coasts of North Africa, and the Moroccan coast in particular. The prince cruised the Moroccan coast on many occasions, stopping for long periods in Tangier where his engagements ranged from political networking to the purely recreational. On one occasion he was sent to Tangier by his father to appoint a consul in the "White City." Other times he went there for spectacular hunting parties, or to pursue ethnographic research (as in the Portuguese islands where he condemned the colonial exploitation of the indigenous people). But there were two trips as part of his research activities that had very different objectives, one in 1894 and most particularly in 1897. That year, as the first *Princesse-Alice* completed its final voyage, the prince laid over in Marseille and Gibraltar before dropping anchor in Rabat, then Mazagan and Safi. It was on this trip that he first used a Gaumont cine camera–an adventure in photography that made him think, and record in his diary: "I took a few photographs in the streets, looking for the best place to try out the cine camera I have on board. It's something I enjoy hugely, always such a thrill." Henceforth it was through the lens of a documentary filmmaker that he observed the local people go about their everyday lives–fishing, buying and selling, and washing then carding raw wool before loading it onto camels. Then came the year 1905 when he had a hand, albeit indirectly and from a distance, in the outcome of the *Coup de Tanger* (Tangier crisis).

LEFT
Kaiser Wilhelm II (1859-1941).

RIGHT-HAND PAGE
Wilhelm II arrives at Tangier on March 31 1905.

FOLLOWING PAGES
Tangier marketplace and view of the German legation, July 1894. Photography by Prince Albert I.

On March 31 that year, with Germany's economic interests under threat from France's growing influence over Morocco, Kaiser Wilhelm II ceremoniously landed in Tangier and proceeded across the city, retinue in tow, to a meeting with Sultan Abdelaziz where he declared himself prepared to take up arms to prevent Morocco from becoming a French protectorate. With France and Germany now at loggerheads, Prince Albert I offered his services as a mediator knowing that his good relations with Berlin and Paris stood him in good stead. He acquitted himself as a true champion of appeasement, delighted that Maurice Rouvier, President of the Council, should choose to enter into negotiations, unlike his more intransigent colleague, French Foreign Minister Théophile Delcassé. A settlement was finally reached at the Algeciras Conference in 1906, with France enjoying the support of the United States, Britain and Italy–a gratifying conclusion for the prince, never mind that certain French and German newspapers said otherwise.

GERMANY, A MANY-FACETED TERRITORY

Germany for the prince meant first and foremost Stuttgart and Baden-Württemberg, dynastic home of his Florentine aunt who married Count Wilhelm of Württemberg, Duke of Urach, in 1867. As evidenced by the prince's notebooks, the Baden-Württemberg region together with Bavaria, was one of his favorite hunting grounds. There was also Berlin, of course, where he went on official visits and for scientific conferences. Note, incidentally, that the prince's campaigns included several German scientists, among them meteorologist Hugo Hergesell. What interests us here, however, is Kiel, a port on the Baltic Sea that became a focus of the prince's activities. He first went there in 1898 in the course of his first Spitsbergen expedition, stopping along the way to take part in the Kiel Regatta, a highlight of the social calendar then in its sixth edition. The second *Princesse-Alice* being unsuitable for racing, the prince concentrated instead on getting to know Kaiser Wilhelm II, finding him much to his liking, unlike the French who were among the Kaiser's sternest critics. This was at the time when all the talk was about the Dreyfus Affair. Prince Albert praised the Kaiser's intelligence and modern policy-making, regarding him as a friend and an advocate for peace despite his unstable behavior. Yet all the warning signs were there, from Wilhelm's conspicuous absence at the opening ceremony of the oceanographic museum to his lack of cooperation in the creation of international maritime regulatory agencies.

ABOVE
The second *Princesse-Alice*.

RIGHT-HAND PAGE
Prince Albert I and Kaiser Wilhelm II deep in discussion on the bridge of the second *Princesse-Alice*, June 28 1907.

FOLLOWING PAGES
Left: Sailor climbing the rigging of the *Meteor*, Kiel, June 1907. Photograph by Justinien Clary.

Right: Kaiser Wilhelm II in sou'wester aboard the *Meteor* imperial racing yacht, Kiel Regatta, June 1907. Photograph by Justinien Clary.

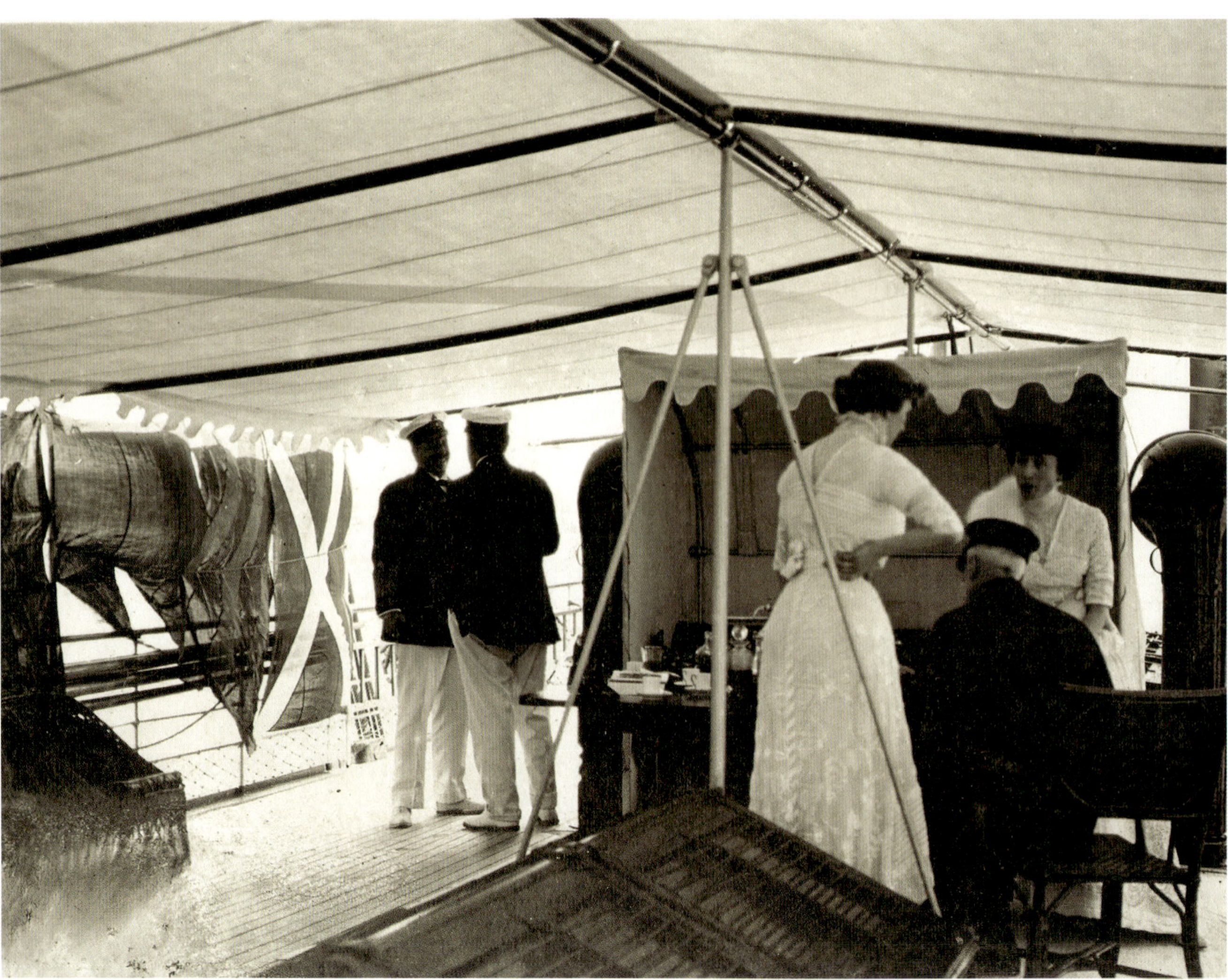

LEFT-HAND PAGE
Top: The prince and his guests aboard his yacht, Kiel, June 1907.

Bottom: Life on board during the Kiel Regatta: Kaiser Wilhelm seen here on the left relaxing with a member of his close entourage.

RIGHT
Crew of the *Meteor*, Kiel, 1907. Photograph by Justinien Clary.

Nothing daunted, the prince put his faith in informal diplomacy when serving as a mediator between France and Germany–two nations with origins and interests in common. He hoped that an exchange of views would be enough to settle the dispute. Going beyond the culture versus civilization concept of his time, in 1903 he dedicated the German version of *La Carrière d'un navigateur* to Wilhelm II, who he introduced as a man of peace and progress. But his illusions vanished when, aboard his imperial yacht the *Meteor* that was taking part in the Kiel Regatta, he learned of the Sarajevo Assassination on June 28 1914. The news followed the prince's ultimate attempt at negotiation, and dispelled any hope of avoiding conflict. A few days later, he began work on a book recording his annual visits to Kiel, *Réflexions sur seize années de visite à Kiel.* With his dream of internationalism in tatters and his hopes bitterly dashed, the prince became fiercely anti-German, venting his anger in *La Guerre allemande et la Conscience universelle*: a virulent denunciation of the Kaiser's duplicity, published in 1919, expressing a rage that may seem surprising in such a moderate man.

DREAMS OF AMERICA

America for Prince Albert I was the complete antithesis of the disappointment that was Germany. He made three visits to America, the first as a young crown prince serving in the Spanish Navy. Even then, his experience of Cuba and Puerto Rico pricked his humanitarian conscience, filling him with dismay at the treatment of the indigenous people by the local landowners: "Slavery has been abolished; but the men born on lands civilized by whites are dispossessed of their heritage by force when deceit fails, their blood already poisoned by the alcohol so skillfully instilled into their veins by colonial expansion." (*La Carrière d'un navigateur*). From there he embarked for the United States, visiting New York and Chicago and the places in between–an experience that taught him a lot about the New World under construction. The memories of that first initiatory trip fill an early chapter of *La Carrière d'un navigateur*.

His second trip was in 1913, a lengthy stay lasting from August 5 to October 28 when he toured North America all the way from the St. Lawrence Estuary in Canada to Wyoming in the western United States. It was a trip particularly rich in symbolism for the prince, bringing him face to face with the wide-open spaces of American mythology but also the uncompromising modernity of the New World. His meetings with Alexander Graham Bell, William Frederick "Buffalo Bill" Cody and President Woodrow Wilson say it all. Staying as the guest of Bell and his wife at their Beinn Bhreagh Estate in Baddeck, the prince evinced the same fascination for the hills of Nova Scotia as for the inventions of his esteemed host–who also died one hundred years ago this year. In Wyoming the prince found himself in a hunting ground par excellence with none other than "Buffalo Bill" as his guide, an adventure they commemorated by naming their campsite "Camp Monaco".

Yellowstone Park, United States
trip, September-October 1913.

But beyond bagging big game, the prince also took pleasure in Nature, in this Wild West still untouched by mankind that reminded him of the ocean wildernesses so dear to his heart. Those wildernesses that were the perfect antidote to a civilization he found too refined for his taste—to those tourist attractions touted ever since the 1850s so much decried by the father of American conservation, John Muir. But here, bathed in the solitude of spectacular landscapes, the prince could reflect not only on himself but also on the need to protect these precious environments. Next stop Chicago, New York and Washington—from landscapes to cityscapes and the metropolis in all of its glorious modernity. Rail networks offering a range of services unparalleled in Europe at the time; illustrious venues where the prince could present his findings to the members of different learned societies; and a wealth of institutions embodying American universalism and industrial prowess. Not forgetting, of course, a meeting with President Wilson that filled the prince with admiration for his interlocutor's internationalism—the attractions of the city were not lost on Prince Albert I. That said, he retained his capacity for critical thinking and looking beyond the obvious. His visit to the famous Chicago slaughterhouses, for instance, found himhorror-struck by the sheer butchery of the killing dictated by the capitalist credo.

BELOW
Left: Albert I and Buffalo Bill set off from Pahaska, September 29 1913.

Right: Albert I and Buffalo Bill in Sheridan Avenue, Cody, Wyoming, September 16-17 1913.

RIGHT-HAND PAGE
Yellowstone Park, September-October 1913.

Following on from that second trip, the prince returned to the USA in 1921 to receive two prestigious decorations: the Cullum Geographical Medal, awarded to him on April 23 by the American Geographical Society of New York, at the Engineers' Club Building; and the Alexander Agassiz Medal, awarded to him on April 26 at the annual banquet of the National Academy of Sciences, at the Powhatan Hotel in Washington. The day before, on April 25, also before the members of the National Academy of Sciences, the prince delivered his famous Speech on the Ocean—a testament to a lifelong endeavor driven by his curiosity for the living world. Saluting the ideals of peace and civilization, the prince also made his now celebrated and pioneering plea for marine conservation. He decried the ill effects of deep-sea and intensive fishing, already warning of species extinction and bringing attention to the plight of wildlife but also the fragility of plant life, a vital resource whose survival depended on concerted action at international level and "uniting humanity in pursuit of justice, jobs and freedom."

LEFT-HAND PAGE
Life in the great outdoors, Shoshone National Forest, north of Pahaska, Wyoming (Camp Monaco).

BELOW
At the end of a black bear hunting trip; photo shoot, October 10 1913.

Camp Monaco

Between August 5 and October 28 1913, Albert I toured Canada and the USA. It was on this occasion that he met up with William F. Cody, alias Buffalo Bill, at Camp Monaco in Yellowstone National Park (Wyoming) and embarked on a hunting trip that brought home to him the importance of national parks and nature conservation.

LEFT-HAND PAGE
A day at Camp Monaco, October 1913.

RIGHT
The prince at Camp Monaco, Yellowstone National Park, Wyoming, on an excursion in the Yellowstone Rocky Mountains, September 29–October 6, 1913.
From left to right: Ferdinand Loüet, Abraham Archibald Anderson, Prince Albert I, Henry Bourée and Louis Tinayre.

JOURNEYS TO ITALY

No account of the life of a gentleman would be complete without mentioning the ritual journey to Italy. For Prince Albert I of course, Italy was not one but several. There was the Italy closest to home, the Italy of the Balzi Rossi caves where he supported the archaeological excavations led by Canon Léonce de Villeneuve exploring the Grotte du Prince (1895), the Grotte des Enfants (1900), the Grotte du Cavillon (1902) and the Abri Lorenzi (1914). There was the Italy of Classical History, Rome and the islands of Italy that in 1893 wooed Princess Alice onto her husband's boat, then on an expedition in the Mediterranean, to satisfy her taste for the Antique. Putting his research (and the objections of his scientists) to one side, the prince landed in Naples, Capri, Syracuse and Palermo. Then there was the Italy of the Holy See–the Vatican City with which the prince, out of respect for tradition but with little inclination for things clerical, entertained a complicated relationship that tested diplomacy to its limits. Suffice to mention the incident in 1904 when, following President Émile Loubet's visit to King Victor-Emmanuel III of Italy, the prince's entourage sent a note to the French daily newspaper *L'Humanité* expressing the pope's objections. The ensuing outcry would have repercussions on the 1905 French Law on the Separation of the Churches and the State.

Villa Adriana, Tivoli, Italy, April 26 1910: the prince with Queen Elena of Italy and her sister Vera del Montenegro. The following day the prince delivered a lecture to the Italian Geographical Society in Rome.

Our story of the prince's passage through life ends here. Not so much an exhaustive account as a call to the imagination, an invitation to look beyond the clichés (Scholar, Navigator, Humanist) and absorb the prince's oeuvre in all of its impressive complexity. An account that we hope will inspire others to prolong the adventure–delve into the document corpus presented here to chart a different course (reconnoiter other paths, slightly further east, headed perhaps for Romania or Imperial Russia). Historical records and illustrations alike testify to a man whose sociability could have surprising consequences. A man who was equally interested in many things, embodying the curiosities and paradoxes of his age but also caught between a fascination with progress, persistent doubt about the pitfalls facing civilization, and sober reflections on the plight of nature. A man driven by a vision whose philosophy and activities took many forms, and whose involvement in the affairs of his time, whether or not crowned with success, must serve as a lesson for the present.

ABOVE
The prince sets up his camera in the ancient ruins of Villa Adriana, Tivoli, April 26 1910.

RIGHT-HAND PAGE
Prince Albert I standing alongside a winged victory statue of the Goddess Minerva, ruins of Ostia Antica, Italy, April 28, 1910.

"These ruins, like so many others in Rome, leave a vivid impression of the greatness of things [...] I wandered around them for hours, every temple, every basilica and every palace filled with a magnificent silence teeming with the secrets of the life they once beheld."

Albert I of Monaco, Rome, April 1910.

Photo session, between Castel Porziano and Ostia, April 28 1910.

Harris & Ewing
HE
WASHINGTON D.C.

"In the midst of the elements that together form the harmony of our earthly globe, marine plants often play the role of intermediaries between the extinct and the living world on the Earth. Sadly, many of these plants too often go to waste in some places. It is as though man cast prudence to the wind at the sight of Earth's riches. Succumbed to a dizziness that led him to destroy everything around him, for there is no natural product on Earth that can survive the careless handiwork of human industry."

Speech on the Ocean by Prince Albert I, delivered at the National Museum in Washington to the members of the US National Academy of Sciences, April 25 1921.

Photograph of the prince taken by Harris & Ewing in Washington D.C., April 1920.

Tributes

to Prince Albert I offered by the Literary Council of
the Prince Pierre of Monaco Foundation

To the memory of academician Prince Albert I of Monaco

——— by **XAVIER DARCOS**

of the Académie Française, member of the Academy of Moral and Political Sciences, Chancellor of the Institut de France

It was a moment filled with emotion. That moment on January 23 1911 when Prince Albert I of Monaco officially opened the Institut Océanographique on the Montagne Sainte-Geneviève in Paris. Moments later he took the floor to address the then President of the French Republic, Armand Fallières, together with France's foremost state and scientific authorities: "[...] I stand before you, wearing the uniform of the Institut de France that identifies the workers of France's new aristocracy [...]." Prince Albert was indeed a member of the Institut de France having joined the French Academy of Sciences, first as a Corresponding Member then as an Associate Member. So it was in the iconic *habit vert* of the Institut's five academies that Albert I of Monaco went on to give his now famous speech. A magnificent speech, worth reading and rereading today, in which he explains the meaning of his drive to "understand the Sea and deliver its kingdom unto Science."

All of the prince's scientific and cultural activities were perfectly aligned with the spirit of the Institut de France. But above all, his endeavors were a shining testament to a man whose commitment to exploring a little-known realm achieved, to use his own words, "a conquest that shed a new and brilliant light on the mysteries of the Deep." What is more, his was a collaborative undertaking: "conquering the unknown through knowledge" was a concerted action conceived and implemented alongside his fellow scientists. Then there was the prince's determination to publicize the findings of his ocean research, or what he described as "those truths that every citizen has the right to know in order to acquire the serenity that conquers passions."

There is no more relevant message for our time than this commitment to promote the spread of knowledge. In truth, Prince Albert I did not so much wear the *habit vert* as become the living embodiment of the values it represents: scientific advancement; collaborative action research; and the sharing of knowledge for the benefit of all. He was the very epitome of that "new aristocracy" he saw reflected in an academician.

At the prince's funeral on July 8 1922 in Monaco, Louis Joubin, speaking on behalf of the Academy of Sciences, paid tribute to the deceased's assiduous attendance at the academy meetings, for as long as his declining health would permit. Above all, Mr. Joubin commemorated the prince's scientific oeuvre, describing it as "so diverse yet so beautifully unified, so great and yet so simple." A lifetime dedicated to the sea "to which he was drawn by his appetite for independence and the poetic longings aroused by Nature's immensity."

Sailors have always scoured the horizon. This sailor prince saw so far beyond the horizon that it was not enough merely to champion the arts and culture. Instead, he focused his energies on what we now call "environmental" issues long before they became an issue. Creative, committed and persevering, Prince Albert I of Monaco was a true man of science. But he was above all a visionary. And it is this visionary quality that touches us academicians most here. A visionary quality that made him such a credit to the *habit vert* and his fellow members of the Institut.

A Man amongst men

by **FRANÇOIS DEBLUË**
Francophone writer representing the Swiss French-speaking literature

There was a prince
who was a real prince:
a lover of justice and science
a philosopher sailor
and a man amongst men

Come Hell or high water
the oceans were his passion:
his was a tireless quest
to plumb their depths
and delve into their beauties

Of another Captain - Dreyfus!
he took up the defense
- and with it, the defense of a defenseless peace

To courtiers and honors
he would have preferred
Humanity's truthfulness
- as such his greatest riches were the wealth
of his soul and spirit.

February 2021

Sovereign, Nomad, Explorer of Continents

by **MARIE-CLAIRE BLAIS**

Francophone writer representing the Canadian French-speaking literature

As a child, then a teenager, he could already see his dream taking shape in the distance,

He would rule in pursuit of peace and justice,

He would be curious about the World's peoples and ever respectful of their traditions,

He would share, not dominate, because domination only succeeds through love,

Unencumbered by the lust for power, his ideas quickly made of him a poet builder and inventor

A serene man, the world in all its splendor awaiting discovery, sea-tossed ships caught in the sun

In the silence of the night amid the whistling waters he dreamed of departure, of being everywhere

Where a humanity as yet unknown lay simmering, veiled by expanses of water, sea and ocean. Pedaling fast beneath palm trees, he leaned sideways with an airy soul, breathing in

The perfumed air and seeming about to take flight as he danced on the balmy air, borne on the waves of birdsong

He was free, free to break the chains of bondage and restore to the people the freedom that was their due, he knew the taste and the thrill of freedom, knew just how vast and complex was this world

That still held marvels in store, he was a young prince longing for knowledge and discovery, far removed from any wish for domination, as a child then a teenager,

He could already see his dream taking shape in the distance, see himself sailing away, hair blowing in the wind, on the ships and vessels that he so skillfully painted, he had to be reminded to behave, to curb his dreams, to remember that he was first and foremost a prince who had to learn the art of ruling,

With a warm and tender gaze, he envisioned a world without borders, as if he were already captain of his ship,

And then came the day when he was ready to set sail, on September 12 1870, Prince Albert boarded

A sailboat freshly arrived from Cherbourg and headed for Monaco,

On October 14 1873 Prince Albert bought a schooner that he named *L'Hirondelle*–"swallow", a symbol of freedom, its winsome wings like the frail sails of a sailboat. *L'Hirondelle*

Would soar above the water in the teeth of storms and cyclones, revel in movement and adventure as she plowed the airy softness of wave-hugging clouds, gliding gracefully above the surface, Prince Albert felt a vision take shape, he would film his life as it happened, packed with live action and vivid footage of the journeys he charted from one ocean to another, he would keep a journal, record his crossings of the seas and oceans in minuscule detail,

Tomorrow's future was built on knowledge,

Tomorrow's generation would see his pictures but so too would his present-day audiences,

We would all see his films and never forget them, those films from the Institut Océanographique, they were another of the prince's dreams, those films, old films but still as crisp as the day they were made, and Prince Albert being an artist but a scientist too, art and science being in his view largely inseparable, he built a conference hall where he could continue to testify to the beauty of this world,

Seeing was nothing, perceiving was everything, looking beyond what you saw,

Remembering a glacier in all of its radiant beauty, right down to the smallest detail,

Would the glaciers still be there tomorrow, would our planet still teem with wildlife, would the majesty of Nature untamed survive the onslaught of tomorrow? Ever the optimist, the prince cultivated the joy of creativity, celebrated All Things Art, what could he know, this peaceful and gentle man, about the men of tomorrow?

He sensed there were cruel conflicts in store, that the ever-present threat of war was too often obscured, but the prince was an enlightened man who put his faith in the blossoming of human intelligence and the march of progress, the prince was Modernity's Man,

Filming a glacier one day and landing in Morocco the next, the prince strove to capture the moment in all of its feverish intensity, shoot street scenes in Morocco, surrounded by men, women and children sizzling under the summer sun, then depart with a joyful heart

To rejoin his boat and his staff and set a course for São Miguel, ah, the allure of those islands he wrote in his journal, with hot springs that he wanted to photograph, film from his boat, assailed by sailing enthusiasts wanting to be filmed by the prince as they raced in the choppy waters off the coast of Portugal,

Prince Albert I was the first oceanographer and his ocean discoveries were a rich legacy for the future.

Prince Albert I was also the first Navigator Prince,

His was a life of joyful navigation that he recounted in his journal, filling the pages with notes and scientific observations rigorously penned in his large but careful handwriting that speak above all of his ebullient imagination, every experience recorded to create an unforgettable story of adventure with new happenings at every turn,

The prince's journal, like his life, stands as a testament to his passion for humanity, an often lost and remote humanity that he brought closer to us, insisting that man must be seen in the light of his times however changeable and evanescent those times may be, this journal that was stored in the palace archives to preserve for posterity the precious matter that it contained, the prince's landing in Safi, for instance: the rugged conditions as the boat approaches the cliff; the disembarkation by pirogue; the coast teeming with fascinating fishes that turned out to be mackerel but a marvelous sight nonetheless for those with eyes to see

Soon he would launch his second sailboat, *L'Hirondelle 2*, in Seyne-sur-Mer,

Well, not so much a sailboat, more a long-legged vessel

That would carry him to unknown lands: the kingdom where it all began for this sovereign, nomadic prince who was an explorer of continents.

The melancholic voyager

by **PHILIPPE CLAUDEL** of the Académie Goncourt

I would have liked to read *La Carrière d'un navigateur* as a child because the word "navigator" would have caught my attention as surely as a trawl net, even though I hated the sea and still don't like it very much.

But to navigate is not just to board a vessel, whatever that vessel may be. Rather, it is to venture beyond the horizon. To navigate is to lose oneself in an immensity greater than oneself. To navigate is to sleep under the stars trusting in their benevolent guidance. To navigate is to become Ulysses who delays the nightly return home because the Earth is so big and life is so small. To navigate is to remain forever the child who discovers the world in all of its endlessly renewed expanse.

So yes. I would have read that book and I would have liked it. Like is not quite the right word. I would have been *carried away.*

Carried away by the current, by a burning thirst. Carried away as when I immersed myself in the writings of Jules Verne and Joseph Conrad, Henry Morton Stanley, Lord Carnarvon and Alexandra David-Neel, and became those adventurers when I closed my eyes at night or let my thoughts flow free by day: sailors, explorers, geographers and archaeologists whose every move I observed on every written line.

I am now a few years older than Albert I when he published his book, so it was carrying my life as baggage when I first read the memoirs of this voyager sovereign.

To depart for an elsewhere has the virtue of stripping away the persona—how we see ourselves and how others see us. The king stripped bare amid the infinity of Pascal-like spaces. And so much the better for the king, since what interests us here anyway is the man behind the clothes. When confronted with a harsh and often hostile environment, the outward trappings of wealth and status count for nothing. Empty mumblings, nothing more. Spindly twigs reduced to flying embers and ashes quickly scattered to the four winds.

What remains is the Inner Being, whose sense of wonder at this untamed world so full of riches, so well conveyed in his book, gradually morphs into a melancholic meditation on who we are and the way the world is going, or rather is not going. A point raised by those illustrious predecessors so often referenced by the prince in these pages. One thinks of Rousseau. One thinks of Edgar Allen Poe's *Narrative of Arthur Gordon Pym.* One thinks of Thoreau and of Melville. Presented in simple prose, with no pretensions to being great literature, his writing touches me more than I can say.

Albert hunts. Albert navigates. Albert eats. Albert climbs mountains, makes lists, collects plants, Albert fishes. Albert sleeps. Albert observes his companions. Albert loses himself in landscapes. Albert forgets he is Albert I. Albert has eyes only for the world and the world has eyes only for him.

Albert whittles himself down as the story reaches its climax, so that by the last pages of the book all that remains are the reveries of the symbolically solitary navigator, a wanderer of frozen wastes lamenting the inherent dirtiness of human civilization. A man alone, dwarfed by the immensity of the Arctic who, to quote Rimbaud in *Lettre de Gênes* sees "nothing but whiteness out of a dream, to touch, to see, or not to see?"

The closing lines of the book thus serve as an aleph: the beginning and end of all things–a thought, a life, an experience, this account of an expedition in the Frozen North to the incandescent pallor of Spitsbergen. But they also convey a singular wish that tugs at my heartstrings:

How good it would be to die there, caught between memories of loves lost and cruel separations and dreams of happiness; far from the passions born of humanity's vices.

Suddenly there is no more Albert I, no more prince, no more navigator, no more Albert at all.
There is nothing and nobody left.
There is nothing left but us and what faces us.
Us, still alive, just,
But only just.

February 21 2021

Twenty medals for a prince

—— by **MARC LAMBRON** of the Académie Française

1. There is something about Albert I, Prince of Monaco, that lends itself to fictional treatment. Something straight out of a book or a movie that makes you think of the staccato shots of early cinema and the silent images of abyssal fish dreamed up by Georges Méliès. An airy quality not unlike the composer Ravel as imagined by Jean Echenoz.

2. The prince was born in 1848, the year of European revolts, in the Rue de l'Université in Paris. An auspicious year to say the least. He departed this life in 1922, a few months before a friend of his granddaughter's husband, the writer Marcel Proust. In other words, he lived a long time.

3. It seems that the prince's interest in science started with a desire for the sea. At twenty-two he kitted out his first schooner, The *Hirondelle*, based on a robust hybrid approach that saw him recruit Breton crews for ships built in the Royal Navy Dockyards in England.

4. The prince nurtured the encyclopedic aspirations of the Enlightenment: to place voyages of exploration at the service of science; to promote the creation of a mathesis universalis; to encircle the globe in networks of knowledge. Even Louis XVI himself, shortly before going to the scaffold, reputedly asked "What news of Lapérouse?" The prince understood that fostering the spirit of concord was the best antidote to ignorance.

5. The prince's oceanographic expeditions included a contingent of scientists, recalling the scholars who took ship for Egypt with Napoleon Bonaparte. The prince's scientists had his floating laboratories at their disposal, and included a bacteriologist and zoologist: an accomplished ichthyologist himself, the prince entertained no rosy illusions about the harmlessness of wildlife. But he believed that knowledge illuminates the future just as a beacon guides ships over sandbars.

6. The prince's pelagic expeditions were equipped with a picturesque array of instruments whose names connoted mystery—small trawl nets, beam trawls, echo sounders, bar rigs, triangular traps, plankton pumps and more.

7. The prince was a very good cartographer. In the Azores, he discovered a 3,150 meter-deep trench, and drew up a map of the currents and oceanic bathymetry. Being a lover of Chthonic realms and all things deep and dark, he also visited the Cave of Altamira.

8. In 1910, the prince founded an institute of paleontology. *Homo sapiens* being the very distant descendant of a fish that emerged from the water in prehistory, the study of plankton quite naturally aroused his interest in pithecanthropus.

9. The year 1910 also saw the opening of the Oceanographic Museum of Monaco, established by the prince eleven years earlier as a venue for the presentation of his research findings. For this man driven by altruism and didactic imperatives, promoting research was meaningless without the democratization of knowledge.

10. The prince was a notable champion of progress and, with it, sporting performance. By 1895, eight years before the first Tour de France, he was already riding velocipedes and three-seat tandems. In 1897, two years after the invention of the Lumière Cinématographe, he tried out the first

movie cameras in Morocco. In 1904 he raced the first powerboats; in 1905 he organized Monaco-Paris motorcycle rallies; in 1907 he took to the air in a dirigible; and in 1912 he flew over the principality aboard one of the first hydroplanes.

11. The prince expressed polite curiosity in everything he saw, most particularly in the Azorean women enveloped in their large hooded cloaks, who he apparently found more surprising than the polar bears in Spitsbergen.

12. In 1913 the prince went hunting in America with notorious bison killer William "Buffalo Bill" Cody, whose antics as a hunter may have provided the inspiration for the twists and turns of the Monaco Grand Prix. We shall never know.

13. The prince belonged to many learned societies, among them the Académie des Sciences, which he joined as a Corresponding Member in 1891 before being appointed a Foreign Associate Member in 1909; and the Académie de Medicine that elected him an Associate Member in 1915. It is not known whether any of his fellow members ever met Buffalo Bill.

14. The prince, like the Aediles of ancient Rome, understood that a city needs a museum of Fine Arts quite as much as it needs a waste incinerator. He commissioned the building of both.

15. The prince met at least four presidents of the French Republic: Sadi Carnot, Félix Faure, Armand Fallières and Paul Deschanel. For three of them, their term of office met with an unfortunate, abrupt end. The prince had nothing to do with it.

16. The prince was a staunch defender of wildlife, with a particular fondness for two composers who wrote music about animals: Camille Saint-Saëns, whose "Carnival of Animals" provided the musical theme for the Cannes Film Festival[1]; and Jules Massenet, composer of the opera *Méduse*, the song *L'âme des oiseaux* and incidental music for *Le Grillon du Foyer* (The Cricket on the Hearth) and a play by Victorien Sardou (the piece itself being called *Le Crocodile*).

17. In the name of enlightenment and humanity, the prince showed unflagging support for Captain Dreyfus and his family. His devotion to their cause is especially commendable as few if any of his fellow movers and shakers did the same.

18. The photographs and prints illustrating the prince's life are uncannily reminiscent of a certain famous author. Giving a talk in London or New York, he makes you think of Professor Paganel in *In Search of the Castaways.* Standing squarely on the bridge of his schooners as storm rages around him, he reminds you of the proud sailors in *The Survivors of the Chancellor.* And those bizarre creatures that he brings back from the deep? Monsters straight out of the illustrated edition of *Twenty Thousand Leagues Under the Sea* published by Hetzel in 1871. Pure Jules Verne in other words. Captain Nemo meets Captain Cousteau, with a touch of Indiana Jones' father thrown in for good measure.

19. When you look back at the prince's 33-year reign, what you see is an exemplary ruler who eschewed the sybaritic beach life in favor of leading his people. Someone like 13th-century emperor Frederick II of Hohenstaufen: an accomplished falconer, thoroughbred breeder and linguist; a passionate astronomer and pioneer of human anatomy; in short an enlightened monarch who welcomed Jews and Muslims alike to his court. The prince in his turn may well have provided the inspiration for *Babar the King*, the elephant character invented by Jean de Brunhoff in 1931. King Babar, like the prince, was a pacifist and environmentalist, a ruler who combined wisdom in all things with unfailing equanimity and a commitment to promote harmony between people. Also like the prince, Babar, Paul-Valéry style, knew how to talk with fish under the sea.

20. A quote from the prince: "The servants of science everywhere are today united by the imperious necessity to make up for lost time in progressing toward new ways of thinking and doing that can ensure the defense of reason against the assaults of uncultured instinct." Still relevant today would you say?

A literary hero

by **FRÉDÉRIC VITOUX** of the Académie Française

The "Belle Epoque"! The mere sound of those words brings to mind graceful, corseted creatures in frothy confections of purple and pink who would pose for Boldini, Gervex and Helleu before tarrying a while at the Pré Catelan restaurant or being seen in a box at the Paris Opera before returning home to their *Hôtel Particulier* in the Faubourg Saint-Germain, with the Guermantes for neighbors, say.

Fortunately for us all, the beauty of the Belle Epoque was not confined to flighty frivolities that fled in the face of the "Storm of Steel" as Ernst Jünger so memorably called World War I.

The turn of the 20th century witnessed the greatest scientific breakthroughs in history. Marie Curie discovered X-rays that could penetrate opaque materials; Max Planck and Einstein overturned Newtonian mechanics; and in 1905 Einstein published his Special Theory of Relativity showing that space and time were intimately linked.

Above all, the Belle Epoque was the period that saw the last great explorations. Henceforth what had hitherto been a vague outline on world maps–the heart of Africa, oceanic archipelagos, the Arctic and the Antarctic and especially the North and South Poles–would be reconnoitered, described, and mapped down to the last detail. Starting with the sea, where knowledge barely extended beyond the surface and the major ocean currents. A whole new continent beckoned: the ocean as a whole, complete with its flora, fauna, landforms and seemingly bottomless depths. So was born a new area of science, and with it the vocation of a man who many would know as the "scholar prince" or the "navigator prince": Albert I Prince of Monaco, a pioneer of the newly coined discipline of oceanography.

The real beauty of the Belle Epoque lies in those *extraordinary voyages* of scientific and terrestrial exploration. *Extraordinary voyages.* And there we have it!

The mere mention of those words is enough to whisk me back to the magic of childhood and youth. I must have been among the last of a generation of readers to devour Jules Verne's *Extraordinary Voyages*, the name given to the original Hetzel collection that featured engravings by Riou, Neuville, Ferrat and Benett whose realistic images made the fabulous believable. They were a part of my father's childhood, those collections, and he held onto them into adulthood. Which is how I came to fly for five weeks in a balloon, drop from the sky onto a mysterious island, make a tour of the world in eighty days and explore the mighty Orinoco.

Well, when I think back on the adventures of Albert I, Prince of Monaco, what comes to mind are those Extraordinary Voyages. Or to put it more simply: few real people have ever seemed to me more deserving of fictional status than this explorer prince.

Poring over the photographs taken on board the *Princesse-Alice* and the sailboats that came after her, I think of the illustrations by the prince's artist friend Tynaire who accompanied him on every trip. And that's when there is no longer any doubt in my mind that the prince was one of those extraordinary men who more rightly belong to the realm of fiction. It isn't possible, we say. You couldn't make it up, we say. Ergo it must be fiction and never mind the contradiction.

There he is in his sailor's uniform, the adventurous man of science, his expression softened by a beard and moustache that will soon go gray. Standing on the bridge of his floating laboratories, he radiates the resolve of an intrepid and serene voyager bound for unknown worlds beyond his dreams. He is a man of his century, the best of his century.

Don't get me wrong! I'm not comparing Albert I to Jules Verne—your sedentary writer par excellence, whose sole experience of travel was to move from Nantes, where he was born in 1828, to Amiens, in Picardy, where he wrote most of his books before eventually breathing his last.

Jules Verne was twenty years older than the prince. He could have been his father. But Albert I did far more than a son could ever do. He became one of Verne's heroes, though at the risk of repeating myself, even Jules Verne would have been hard pushed to invent anyone quite as exceptional as Albert I. Here was a man who combined a passion for adventure with a love of learning; a man who put his trust in scientific progress and the march of technology; a reigning prince fiercely attached to his principality, his "rock", yet irresistibly drawn to wide open spaces and boundless depths.

To this must be added a set of convictions that took a lot of political courage in the period 1870-1914. A humanist, pacifist and confirmed Europeanist, Albert I was nothing if not insightful. And that's not all. In his brief moments behind the scenes as the writer of *La Carrière d'un navigateur,* published in 1902, he wrote the memoir of a life, his own, always focusing on the circumstances or the research at hand and never on himself.

I leave it to others better qualified than myself to enumerate Albert I's many contributions to science and oceanography. For my part, I want to return to his fictional quality—to that aspect of him that would have thrilled the adolescent I once was. The teenager who would pore over Jules Verne's *Extraordinary Voyages* and see himself as one of the characters. Imagine himself being cast adrift like Captain Grant's children, joining in the adventures of Captain Hatteras or eagerly stowing away to Propeller Island. The teenager who would have given anything to sneak aboard Captain Nemo's *Nautilus* and venture 20,000 leagues under the sea.

.../...

If only I had known then that Albert I was a superhero come to life! That here was a man who enlisted in the Spanish Navy at the age of 18, then joined the French Navy as a ship-of-the-line lieutenant to take part in the ill-fated war against Prussia in 1870. A man who went on to lead numerous expeditions to the Azores and Spitsbergen; who installed movie cameras on board his ships as early as 1895; who had a deep-sea fish christened after him, *Grimuldichtys profondissimus*; and who developed the first technologies for the exploration of deep-ocean trenches!

True, Albert I didn't build the *Nautilus*–he did something even better. Thanks to him, a sight previously reserved for Captain Nemo's passengers peering through the portholes of his giant submarine suddenly became available to everyone. By designing and building the Monaco Oceanographic Museum, Albert I created a formidable research resource, but also and more importantly, he gave everyone an opportunity to observe underwater fauna. And as if that weren't enough, in 1906 he founded our very own oceanographic institute here in Paris.

Twice a year I get together with my fellow jury members to pick the winners of the Prince Pierre Foundation literary awards. On those occasions I am filled with warmth but mainly undisguised respect for our lady president, HRH the Princess of Hanover. To be the great-great granddaughter of Albert I, a man who could have been one of Jules Verne's most emblematic heroes, strikes me as an honor that to my knowledge no-one else can ever lay claim.

RIGHT-HAND PAGE

"Free Man, you will always cherish the sea."

Charles Baudelaire

A man is born a prince.
But when does a man choose his own path?
Or his real title?
For the young Albert I this was a beautifully rounded word encompassing Earth and Ocean.
Albert I the Navigator. Not so much a kingdom, more an adventure. A plan B for the world.
On the bridge, in the very early morning.
"Transported to fairyland on the wings of dreams ..."

Free Man, you will always cherish the sea

by **DANY LAFERRIÈRE** of the Académie Française

Acknowledgments

I would like first of all to extend my sincere gratitude to H.S.H. Prince Albert II and HSH Princess Caroline of Hanover.

I also extend my warmest thanks to Professor Jean Malaurie for his heartfelt foreword, and to the members of the Literary Council of the Prince Pierre of Monaco Foundation for their meaningful tributes that echo the talented writings of Prince Albert I: Messieurs Philippe Claudel, Xavier Darcos, François Deblüe, Dany Lafferière, Marc Lambron, Frédéric Vitoux, with a special thought for Madame Marie-Claire Blais who departed this world in 2021. My thanks also go to Monsieur Jean-Charles Curau, Secretary General of the Prince Pierre of Monaco Foundation, for his permission to compile this book.

This book is being published as part of the celebrations to mark the centenary of the death of Prince Albert I so I am especially grateful to the Government of the principality, the institutions and organizations represented and all of the members of the Steering Committee and Executive Committee, without whose support this book would not have been possible.

A special word of thanks to Monsieur Robert Fillon, President of the Commemoration Committee, and Monsieur Thomas Fouilleron, Vice-President of the Committee and Director of the Palace Archives and Library of the Prince's Palace, for their unfailing support and warm encouragement.

Thanks should also go to the officers of the heritage funds, particularly in Monaco, who allowed access to the documents presented in this book: the Prince's Palace Archives, the Oceanographic Institute–Prince Albert I Foundation, the Audiovisual Institute of Monaco and the Société des Bains de Mer. Thank you also to their colleagues and all of those people who cooperated and lent friendship and support to this project: Madame Elisabeth Baltzinger, Head of Archives of the Oceanographic Museum; Monsieur Thomas Blanchy, Deputy Director of the Prince's Palace Archives; Monsieur Michaël Bloche, Director of the *Mission de préfiguration des Archives nationales* (National Archives preparatory team); Madame Charlotte Lubert, S.B.M. Heritage Officer; Madame Chloë Raymond, Assistant Curator of the Oceanographic Museum; Monsieur Christian Roti, Graphics and Communication Manager, Audiovisual Institute of Monaco; and Monsieur Vincent Vatrican, Director of the Audiovisual Institute of Monaco.

I extend my special thanks to Madame Elsa Milanesio, Head of Communications, and Monsieur Fabrice Blanchi, Head of Research, for their assistance in designing this book: to Elsa for her help with selecting the book images; and to Fabrice for his help with writing the captions. Thanks also to Pauline Dubuisson and Virginie Mahieux at Editions de La Martinière for their help in designing this book.

Lastly, a special word of thanks to Madame Jacqueline Carpine-Lancre, whose authoritative writings provided the inspiration for Prince Albert I's collective works.

Then there are all of those close friends who supported me throughout this project– my thanks to Léontine, Marie and Sandra. Thanks to Jacqueline and Jean-Marie. Thanks also to Annie, Louis, Marion and Nicolas.

Stéphane Lamotte

Bibliography

Albert I, prince of Monaco, *La Carrière d'un navigateur*, Monaco, Éd. des Archives du Palais princier, 1966, XXII, 240 p.

Albert I, prince of Monaco, *La Guerre allemande et la conscience universelle*, Paris, Payot, 1919, 170 p.

Albert I, prince of Monaco, Jules de Guerne and Jules Richard, *Résultats des campagnes scientifiques accomplies sur son yacht par Albert Ier, prince souverain de Monaco*, Monaco, Imprimerie de Monaco, 1889-1950.

Jean-Rémy Bézias, « La principauté de Monaco, la Méditerranée et la paix sous le règne du prince Albert Ier (1889-1922) », *Cahiers de la Méditerranée*, 2015, n° 91 : *Du pacifisme à la culture de la paix. Les apports des* Peace Studies *à la construction de la paix*, p. 47-58. [on line] 91 | 2015, on line June 1 2016. URL : *https://journals.openedition.org/cdlm/8071* ; DOI : *https://doi.org/10.4000/cdlm.8071*

Christian Carpine, *La Pratique de l'océanographie au temps du Prince Albert Ier*, Monaco, Musée océanographique, 2002, IV, 332 p.

Jacqueline Carpine-Lancre, *Albert Ier, prince de Monaco. Des œuvres de science, de lumière et de paix*, Monaco, Palace of H.S.H the Prince, 1998, 206 p.

Jacqueline Carpine-Lancre, « Le prince Albert Ier de Monaco et la science », *Archives de l'Institut de paléontologie humaine*, vol. 39, 2008, p. 13-26.

Jacqueline Carpine-Lancre and Luiz Vieira Caldas Saldanha, *Dom Carlos I, roi de Portugal ; Albert Ier, prince de Monaco. Souverains océanographes*, Lisbonne, Foundation Calouste Gulbenkian, 1992, 178 p.

Jacqueline Carpine-Lancre, *Albert Ier Prince of Monaco (1848-1922)*, Monaco, EGC, 1998, 32 p.

Jacqueline Carpine-Lancre, Thomas Fouilleron, Vincent Vatrican et Luc Verrier, *Albert Ier en films*, Monaco, Audiovisual Archives of Monaco, 2014, 96 p.

Jacqueline Carpine-Lancre and Thomas Fouilleron, « Albert Ier, prince de Monaco », *Dictionnaire des étrangers qui ont fait la France*, under the direction of Pascal Ory, Paris, Robert Laffont, coll. « Bouquins », 2013, p. 22-24.

Christian Clot, Philippe Thirault and Sandro (illustrations), *Albert I of Monaco. The Explorer Prince*, Grenoble, Glénat, coll. « Explora », 2018, 53 p.

Ludovic de Colleville, *Albert de Monaco intime*, Paris, F. Juven, 1908, 282 p.

Raymond Damien, *Albert Ier, prince souverain de Monaco. Précédé de l'histoire des origines de Monaco et de la dynastie des Grimaldi*, Villemomble, Institute of Valois, 1964, 519 p.

Thomas Fouilleron, *Histoire de Monaco*, book for secondary education, Monaco, Direction of The Éducation nationale, de la Jeunesse et des Sports, France, 2010, 360 p.

Arnaud Hurel, « L'Institut de paléontologie humaine », *La Revue pour l'histoire du CNRS* [On line], 3 | 2000, on line June 20 2007. URL : https://journals.openedition.org/histoire-cnrs/2992 ; DOI : https://doi.org/10.4000/histoire-cnrs.2992

Henry de Lumley and Anna Echassoux, *L'Institut de paléontologie humaine. Fondation Prince Albert Ier de Monaco*, Paris, Éditions du Patrimoine, coll. « Regards… », 2021, 64 p.

Pierre Miquel, *Albert de Monaco. Prince des mers*, Paris, Glénat, coll. « Une vie », 1995, 300 p.

In addition, scientific articles are regularly devoted to Prince Albert I in the issues of *Annales monégasques. Revue d'histoire de Monaco.*

It should also be noted that Prince Albert regularly features in the journal of Monegasque history, Annales Monégasques, most notably in "*1914-1918. Guerre et paix*", a Special Edition focusing on his pacifist commitment (issue number 38, 2014).

Photographic Credits

Front cover: Monaco's Palace Archives - AIM
Back cover: © Oceanographic Institute–Prince Albert I Foundation, Prince of Monaco

p.1: © Arthur Hewitt - Monaco's Palace Archives - AIM / p.11: © Tallandier/Bridgeman Images / p.13: © Jean-Baptiste Leroux / p.14: © Reserved rights - Monaco's Palace Archives - AIM / p.15 (top): © Monte-Carlo Archives S.B.M / p.15 (center and bottom): © Monte-Carlo Archives S.B.M / p.16: © Reserved rights - Monaco's Palace Archives - AIM / p.19: © Monaco's Palace Archives - AIM / p.20-21: © Monte-Carlo Archives S.B.M / p.22: © Reserved rights - Monaco's Palace Archives - AIM / p.24-25: © akg-images / p.26: © Oceanographic Institute–Prince Albert I Foundation, Prince of Monaco / p.27: © Rue des Archives/PVDE / p.28: © Augustin Aimé Joseph Le Jeune - Monaco's Palace Archives - AIM / p.29: © CCØ Paris Musées/Musée Carnavalet / p.30: © Photo Josse/Leemage / p.31 (top): © Photo 12/Alamy/Yogi Black / p.31 (bottom): © Photo 12/Alamy/Glasshouse Images / p.32 (left): © Photo 12/Alamy/Revival the past / p.32 (right): © Photo 12/Alamy/Michel & Gabrielle Therin-Weise / p.33: © Collection IM/KHARBINE TAPABOR / p.34: © Charles Chusseau-Flaviens - Monaco's Palace Archives - AIM / p.35 (left): © Monaco's Palace Archives - AIM / p.35 (right and top), p.36 (bottom left and bottom and right): © Reserved rights - Monaco's Palace Archives - AIM / p.37: © Jean-Paul Dumontier/LA COLLECTION / p.38: © Rue des Archives/PVDE / p.39: © Coll. Perrin/KHARBINE-TAPABOR / p.40: © Reserved rights - Monaco's Palace Archives - AIM / p.42 (top and bottom): © Reserved rights - Monaco's Palace Archives - AIM / p.43 (top): © Charles Trampus - Monaco's Palace Archives - AIM / p.43 (center and bottom): © Charles Chusseau-Flaviens - Monaco's Palace Archives - AIM / p.44: © Rue des Archives/PVDE / p.46: © Monte-Carlo Archives S.B.M / p.48 (left): © D.R. / p.48 (right): © Bibliothèque Nationale de France / p.49 (top): © Paul Boyer - Monaco's Palace Archives - AIM / p.49 (bottom): © Monte-Carlo Archives S.B.M / p.50 (top): © Bibliothèque nationale de France / p.50 (bottom): © Oceanographic Institute–Prince Albert I Foundation, Prince of Monaco / p.51 (top): © Monte-Carlo Archives S.B.M / p.51 (center): © Reserved rights - Monaco's Palace Archives - AIM / p.51 (bottom): © Monte-Carlo Archives S.B.M / p.52 and 53: © Oceanographic Institute–Prince Albert I Foundation, Prince of Monaco / p.55: © Charles Chusseau-Flaviens - Monaco's Palace Archives - AIM / p.56-57: © Photo 12/Alamy/Bygone Collection / p.58-59: © Granger/Collection Christophel / p.61: © Monaco's Palace Archives / p.62-63: © Photo 12/Alamy/VintageMedStock / p.63 (right): © D. Carlos I Oceanographic Museum Library, Vasco da Gama Aquarium / p.64: © Reserved rights - Monaco's Palace Archives - AIM / p.65 (top): © adoc-photos / p.65 (bottom left and bottom right): © Reserved rights - Monaco's Palace Archives - AIM / p.67 (top, center and bottom): © Oceanographic Institute–Prince Albert I Foundation, Prince of Monaco / p.68 (top): © Photo 12/Alamy/PFRCF Collection / p.68 (bottom): © Granger/Collection Christophel / p.69: © Jean Baptiste François - René Koehler, Public domain, via Wikimedia Commons / p.70 (left): © Bibliothèque Nationale de France / p.70 (right): © Reserved rights - Monaco's Palace Archives - AIM / p.71: © coll. IMI / Agence Martienne / p.73 (top and bottom), p.74: © Oceanographic Institute–Prince Albert I Foundation, Prince of Monaco / p.76: © Photo 12/Alamy/World History Archive / p.77: © Oceanographic Institute–Prince Albert I Foundation, Prince of Monaco / p.78: © Reserved rights - Monaco's Palace Archives - AIM / p.79: © akg-images/André Held / p.80: © Photo 12/Alamy/Hamza Khan / p.82, p.83 (top): © Oceanographic Institute–Prince Albert I Foundation, Prince of Monaco / p.83 (bottom): © Wikimedia Commons / p.84-85, p.86-87: © Oceanographic Institute–Prince Albert I Foundation, Prince of Monaco / p.88: © Monaco's Palace Archives - AIM / p.90: © Oceanographic Institute–Prince Albert I Foundation, Prince of Monaco / p.91: © Norwegian Polar Institute/ Reserve Rights / p.93 (top and bottom): © Monaco's Palace Archives - AIM / p.94, p.95: © Oceanographic Institute–Prince Albert I Foundation, Prince of Monaco / p.96: © Jules Richard - Monaco's Palace Archives - AIM / p.97: © Reserved rights - Monaco's Palace Archives - AIM / p.98-99: © Oceanographic Institute–Prince Albert I Foundation, Prince of Monaco / p.100, p.101: © Monaco's Palace Archives - AIM / p.102, 103, 104, p.105 (top left, top right and bottom), p.106: © Oceanographic Institute–Prince Albert I Foundation, Prince of Monaco / p.107: © Monaco's Palace Archives - AIM / p.108, 109: © Oceanographic Institute–Prince Albert I Foundation, Prince of Monaco / p.110-111: © Photo 12/Alamy/Niday Picture Library / p.112-113: © Rue des Archives/PVDE / p.114: © TopFoto/Roger-Viollet / p.115: © Oceanographic Institute–Prince Albert I Foundation, Prince of Monaco / p.116: © Photo 12/Alamy/Quagga Media / p.117: © SZ Photo / Bridgeman Images / p.118-119, p.120: © Reserved rights - Monaco's Palace Archives - AIM / p.121: © Charles Chusseau-Flaviens - Monaco's Palace Archives - AIM / p.122, p.123, p.124 (top and bottom), p.125: © Reserved rights - Monaco's Palace Archives - AIM / p.126-127: © Ned Warst Frost - Fred J. Richard - Monaco's Palace Archives - AIM / p.128 (left): © Photo 12/Alamy/Imago History Collection / p.128 (right): © Buffalo Bill Historical Center - Monaco's Palace Archives - IA / p.129, p.130: © Reserved rights - Monaco's Palace Archives - AIM / p.131, p.133: © Jack Richard Studio - Monaco's Palace Archives - AIM / p.132 (top and bottom): © Reserved rights - Monaco's Palace Archives - AIM / p.135, p.136, p.137, p.138-139: © Reserved rights - Monaco's Palace Archives - AIM / p.140: © Harris & Ewing - Monaco's Palace Archives - AIM / p.155: © Dany Laferrière

Texts

Stéphane Lamotte

Stéphane Lamotte is an associate professor of history and secretary of the 2022 Albert I Commemoration Committee. He is also the project manager and coordinator of Monaco's Department of National Education, Youth and Sport (DENJS); a lecturer in history, geography and drama; and a research associate and former lecturer at the CMCC (modern and contemporary Mediterranean center) of the University of Nice Sophia-Antipolis.

Graphic design and artwork

Laurence Maillet

Editing

Virginie Mahieux and Pauline Dubuisson

Translation

Florence Brutton

Printed and bound in Printer Portuguesa 10 9 8 7 6 5 4 3 2 1 Abrams books are available at special discounts when purchased in quantity for premiums and promotions as well as fundraising or educational use. Special editions can also be created to specification. For details, contact specialsales@abramsbooks.com or the address below.

Engraving: Quadrilaser

Printed and bound in February 2022 by Grafiche AZ in Italy

Legal deposit: march 2022

ISBN: 978-1-4197-6103-4